PSpice Power Electronic and Power Circuit Simulation

Practical Examples Using the PSpice A/D Demo to Simulate Power Electronic and Electrical Power Circuits

for Electrical Engineers and Technicians

Stephen P. Tubbs, P.E.
formerly of the
Pennsylvania State University,
currently an
industrial consultant

1344 Firwood Dr.
Pittsburgh, PA 15243

NOTICE TO THE READER

The author does not warrant or guarantee any of the products, equipment, or programs described herein or accept liability for any damages resulting from their use.

The reader is warned that electricity and the construction of electrical equipment are dangerous. It is the responsibility of the reader to use common sense and safe electrical and mechanical practices.

Printed in the United States of America and United Kingdom

ISBN 978-0-9659446-9-4

CONTENTS

INTRODUCTION

PSpice is usually used in the analysis of low power electronic circuits. However, it also has the capability of solving power electronic and electrical power problems. This book shows how to use PSpice to quickly analyze common power electronic and power circuits. All examples in this book were done with the Demo Release 16.0 of Cadence Orcad PSpice.

Circuits are given in example problems. In each problem there is a description of the problem, a circuit diagram, a PSpice netlist (PSpice statement list program), and results.

In one example, an engineer wants to evaluate the effect of a notched voltage being transmitted on his mine's power bus. The cause of the distortion had been traced to a power electronic load on the power bus. PSpice would allow the engineer to see how the voltage distortion is transmitted to other loads and to try various corrective circuit additions. This example is in Section 4.1.

In another example, an engineer wants to predict the effect of a line-to-line fault on a three-phase power system. This could be solved manually with the symmetrical components method, but with PSpice, it would be possible to solve many circuit variations faster. Also with PSpice, the three-phase supply does not have to be a pure sine wave. These examples are in Sections 5.4 and 5.5.

This is not a power electronics or power circuits book. It does not cover these subjects in detail or demonstrate optimally designed circuits. It does show how PSpice can be used on power electronic circuits and power circuits. Probably the user of this book will not have circuit problems that exactly correspond to the book's examples. However, with circuit modifications, component value changes, and corresponding netlist modifications, the user should be able to quickly analyze his circuits.

The person using this book should already have an analytical electrical background. Academically, he should be educated to at least the level of a university two-year electrical engineering technology program.

Stephen P. Tubbs

1.0 WHAT IS PSPICE?

PSpice is the most popular **P**ersonal computer version of Spice. Spice is a **S**imulation **P**rogram with **I**ntegrated **C**ircuit **E**mphasis. Spice was designed to simulate electronic circuits on main frame computers. PSpice does the same thing on personal computers.

In the late 1960s, a program called ECAP, **E**lectronic **C**ircuit **A**nalysis **P**rogram, was developed for use on IBM main frames. ECAP was the first program of its type. ECAP and similar programs served as the basis for Spice. In 1984, MicroSim introduced the first version of PSpice. Since 1984, there have been many upgrades to PSpice, and the ownership of the program has changed several times. Now PSpice is owned by Cadence Design Systems Inc.

This book was written using PSpice Release 16.0.

PSpice converts circuits to systems of nodal equations. The equations are then solved numerically. In transient analysis, PSpice uses various numerical integration methods to solve systems of differential equations, limiting the solution accuracy. Occasionally, PSpice cannot solve a circuit's differential equations, and it will produce a "Convergence problem in transient analysis" error message.

2.0 PSPICE AND POPULAR ALTERNATIVES TO PSPICE

2.1 PSPICE

PSpice A/D simulates analog and digital circuits (A/D stands for **A**nalog/**D**igital). It comes integrated with Orcad Capture, a schematic capture program. To simplify the writing and reading of this book, hereafter it will simply be referred to as PSpice.

PSpice's list price in 2008, including maintenance, is $8,995.00.

There are products available as add-on options to PSpice. These are the PSpice Smoke Option, PSpice Advanced Optimizer Option, PSpice Advanced Analysis Option, and SLPS Interface (**S**imu**L**ink **PS**pice Interface).

There are two free versions of PSpice: an Evaluation version and a Demo version. The Evaluation version is fully featured with only a couple weeks of activation. It is distributed to help potential customers decide if they want to purchase PSpice. The Demo version is sometimes called a Student version, Academic version, or PSpice Lite. The Demo version is available for use by those wishing to learn the program. The Demo version handles a maximum of 64 nodes, unlike the full featured version that has no node limit. It also has other reduced capabilities. See Section 7.3.

All of this book's examples were done with the latest Cadence Orcad Release 16.0 PSpice Demo.

The Demo version can be obtained from EMA Design Automation, the company Cadence has contracted to sell PSpice. EMA is located at 225 Tech Park Drive, Rochester, New York 14623. They can be reached at 877-362-3321 and are on the web at www.ema-eda.com. They would be happy to mail a free Demo CD. The CD is loaded with almost 719 MB. Installation of the Demo is mostly automatic, just follow the on-screen instructions.

Older releases of PSpice can be found on the internet. An older computer may not work with the current release, but may operate with an older release. The examples in this book can be done with PSpice releases of more than 10 years ago.

2.2 SPICE AND OTHER SPICES

There are many other circuit analysis programs that were derived from the mainframe Spice program. These can be found with an internet search. One good example is TopSPICE by Penzar. TopSPICE will accept the same netlists as PSpice. Information on TopSPICE and a free demo download is available on www.penzar.com.

2.3 ETAP AND SKM

The ETAP (**E**lectrical **T**ransient **A**nalyzer **P**rogram) and SKM programs are designed to simulate and analyze traditional electrical power generation and distribution systems in normal operation and in short-circuit. They have many add-on options like protective device coordination, arc flash analysis, transmission line sag & tension, raceway calculations, motor starting, etc. They cannot handle the variations of voltages and components that the PSpice and other Spice programs can. For example, PSpice would have no difficulty analyzing a circuit powered by a notched sine wave voltage or powering an unbalanced load. However, ETAP and SKM are only designed for sinusoidal AC, DC, and some common non-sinusoidal waveforms. They will analyze unbalanced faults, but would have difficulty with unbalanced loads.

2.4 MATLAB/SIMULINK

Matlab and Simulink are not circuit analysis programs. They are general use computer programs for solving engineering and scientific problems. However, they can be used to analyze simpler circuits, like those of this book. MathWorks, Inc. makes the programs. Mat is short for **mat**rix and lab is short for **lab**oratory. They are especially good at solving systems of differential equations. Simulink is integrated with Matlab. Simulink creates a graphical environment to make it easier to visualize the problem being solved. It creates a diagram from standard and customizable function blocks. Some manufacturers use Matlab and Simulink when designing power electronic systems that power unstable loads. An example where Matlab and Simulink could be useful would be the analysis of an inverter driving a motor that drives a reciprocating compressor.

There is software that allows PSpice to communicate with Simulink. This is called the SLPS Interface (**S**imu**L**ink **PS**pice Interface). However, the use of it is beyond the scope of this book.

3.0 BRIEF REVIEW OF PSPICE

Example problems are given to refresh the memory of those who have used PSpice or a similar program before. Those who have never done so should study a basic PSpice book, such as those in the references of Section 6.1.

There are two ways of entering circuit data into PSpice. It can be entered manually as a netlist program. A netlist is a list of statements that describe the circuit and the desired output. Alternatively, circuit data can be entered with its schematic diagram with the Orcad Capture program. Orcad Capture then creates a netlist. Currently, the Cadence company promotes the use of its Orcad Capture program.

This book uses only manually entered netlists. Manually entered PSpice netlists are easier to adapt to other types of Spice programs than Orcad Capture schematic diagrams. The schematic capture programs of different Spice programs are usually not compatible. Manual netlist entry requires less learning time. Finally, parameters in manually entered netlists are often easier to modify. This can be advantageous with repeatedly run simulations.

3.1 STEADY-STATE DC CIRCUIT ANALYSIS

PROBLEM: A 10 V_{DC} supply is connected to a 9 Ω resistor in series with a 1 Ω resistor. What is the voltage across the 9 Ω resistor?

PROCEDURE:

1) Draw the schematic and number the nodes.

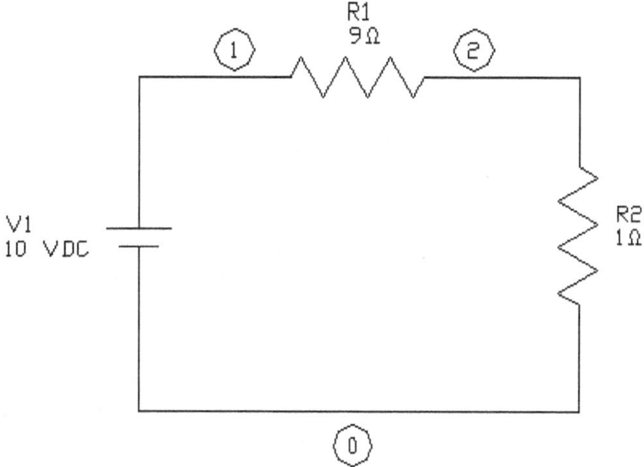

Figure 3-1-1 DC supply with a voltage divider.

Description	Symbol	Value and units
DC supply voltage	V1	10 V_{DC}
Resistance	R1	9 Ω
Resistance	R2	1 Ω

2) Start Microsoft Notepad and enter the following netlist.

```
Figure 3-1-1.CIR RESISTOR DIVIDER WITH DC INPUT
* Finding voltage across R1
V1  1  0
.DC  V1  10  10  1
R1  1  2  9
R2  2  0  1
.PRINT  DC  V(1,2)
.END
```

3) Save the netlist in the directory of your choice as "Figure 3-1-1.CIR". Notepad will save it as a text file that can be read by PSpice. (Note that the Orcad text editor of PSpice Release 16.0 could be used to create the netlist. However, some older releases of PSpice cannot create netlists.)

4) Line by line description of the netlist program:

Line #1, "Figure 3-1-1.CIR RESISTOR DIVIDER WITH DC INPUT", is its title. PSpice expects to see a title on the first line. The "Figure 3-1-1.CIR" is also the netlist's file name. PSpice does not require the file name in its title, but putting it there makes it easier keep files organized.

Line #2, asterisk, shows that this is a comment line. Anything can be written after it without affecting the netlist program. It is a good idea to put the purpose of the simulation here.

Line #3, "V1 1 0", describes the voltage supply. The "1" on the left shows the location of the positive voltage is at node 1. The "0" shows the location of the negative voltage is at node 0.

Line #4, ".DC V1 10 10 1", shows that the V1 voltage is the sweep voltage. The left most "10" shows the start voltage across V1 is 10 volts. The next "10" to the right shows that the stop voltage is 10 volts. The "1" shows that the step value of V1 will be 1 volt. Since the start and stop values are the same, only one value of V1 will be evaluated.

Line #5, "R1 1 2 9", describes the resistor between node 1 and node 2. The "9" shows that the resistor's resistance is 9 Ω.

Line #6, "R2 2 0 1", describes the resistor between node 2 and node 0. The "1" shows that the resistor's resistance is 1 Ω.

Line #7, ".PRINT DC V(1,2)", shows that the voltage across the 9 Ω resistor, V(1,2), should be printed in the output file.

Line #8, ".END", is a required statement at the end of each PSpice netlist.

5) Start the PSpice Demo.

Figure 3-1-2 Opening screen of PSpice Demo. In this screen display, under "View", "Output Window" and "Simulation Status Window" have been toggled on.

6) Select "File" on the main toolbar and in it select "Open".

7) Go to the directory where "Figure 3-1-1.CIR" was placed and open it. The result is shown in Figure 3-1-3.

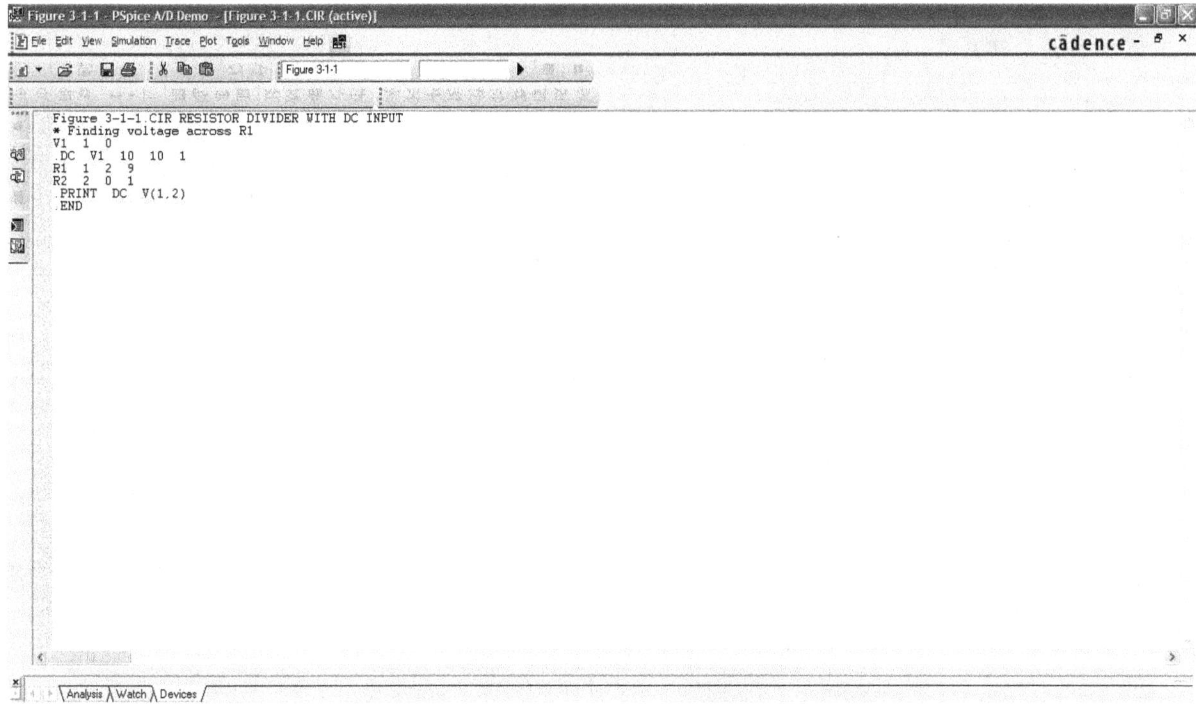

Figure 3-1-3 PSpice Demo with the Figure 3-1-1 netlist loaded into it.

8) In the main toolbar go to “Simulation” and under that select “Run Figure 3-1-1”.

9) The bottom left window should now have in it:

```
--------------- Simulation Circuit File: Figure 3-1-1 ---------------
Simulation running...
Figure 3-1-1.CIR RESISTOR DIVIDER WITH DC INPUT
Reading and checking circuit
Circuit read in and checked, no errors
Calculating bias point
Bias point calculated
Simulation complete
```

10) The bottom right window with the “Analysis” tab left clicked should now have in it:

```
Resistors:        2
Voltage Sour...   1
```

11) In the main toolbar go to “View” and under that select “Output File”. The result is in Figure 3-1-4

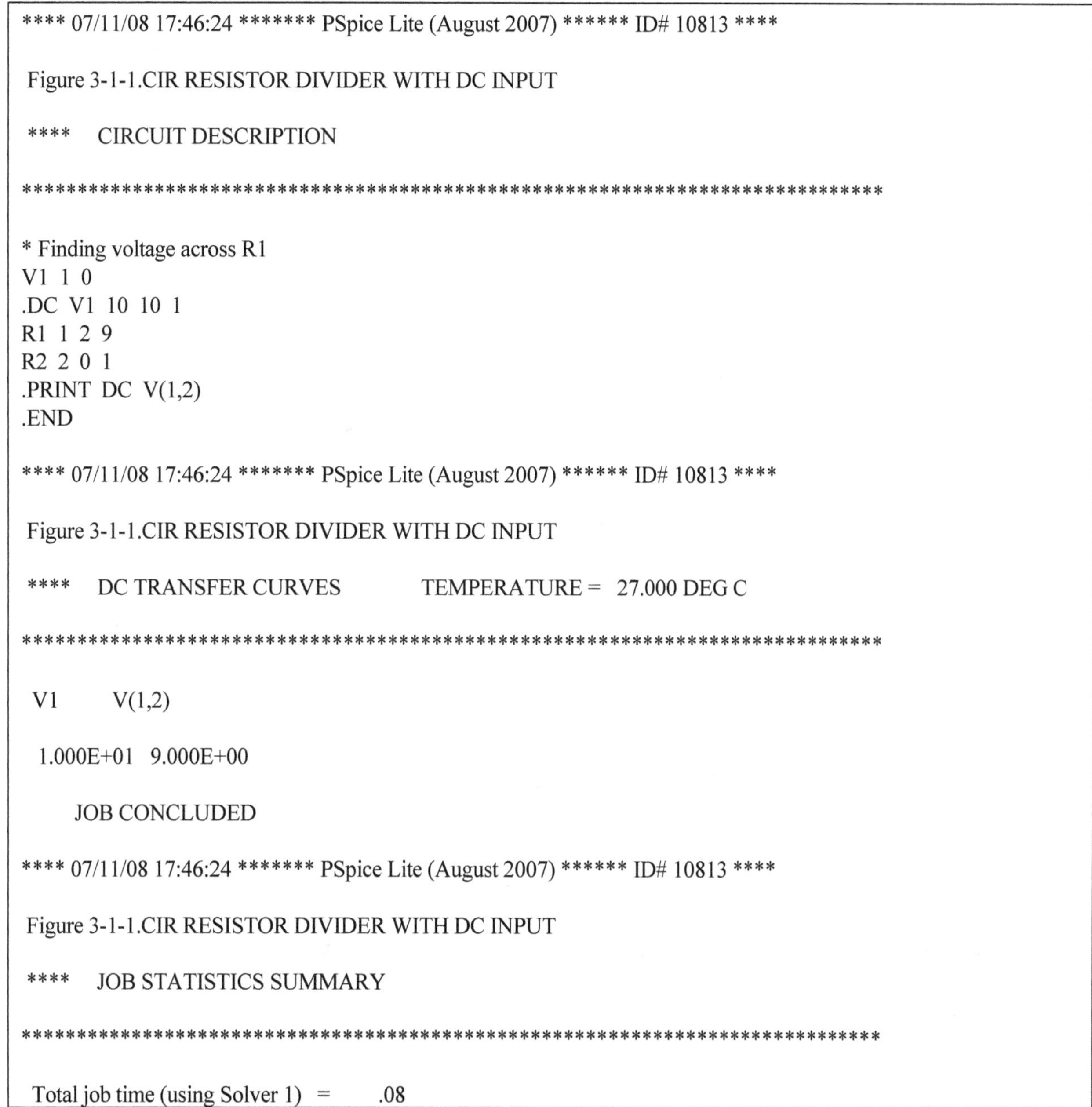

```
**** 07/11/08 17:46:24 ******* PSpice Lite (August 2007) ****** ID# 10813 ****

 Figure 3-1-1.CIR RESISTOR DIVIDER WITH DC INPUT

 ****     CIRCUIT DESCRIPTION

******************************************************************************

* Finding voltage across R1
V1  1  0
.DC  V1  10  10  1
R1  1  2  9
R2  2  0  1
.PRINT  DC  V(1,2)
.END

**** 07/11/08 17:46:24 ******* PSpice Lite (August 2007) ****** ID# 10813 ****

 Figure 3-1-1.CIR RESISTOR DIVIDER WITH DC INPUT

 ****     DC TRANSFER CURVES               TEMPERATURE =   27.000 DEG C

******************************************************************************

  V1        V(1,2)

   1.000E+01   9.000E+00

          JOB CONCLUDED

**** 07/11/08 17:46:24 ******* PSpice Lite (August 2007) ****** ID# 10813 ****

 Figure 3-1-1.CIR RESISTOR DIVIDER WITH DC INPUT

 ****     JOB STATISTICS SUMMARY

******************************************************************************

  Total job time (using Solver 1)   =        .08
```

Figure 3-1-4 PSpice output file from running the netlist of Figure 3-1-1.

12) The "DC TRANSFER CURVES" section of the output shows that the voltage V(1,2) across R1, voltage from node 2 to node 1, is 9.0000 V_{DC}.

3.2 STEADY-STATE AC CIRCUIT ANALYSIS

PROBLEM: A 120 V_{ACrms} 60 Hz supply is connected to a .027 H inductor in series with a 10 Ω resistor. What is the voltage across the 10 Ω resistor? Solve this with PSpice's steady-state AC capability.

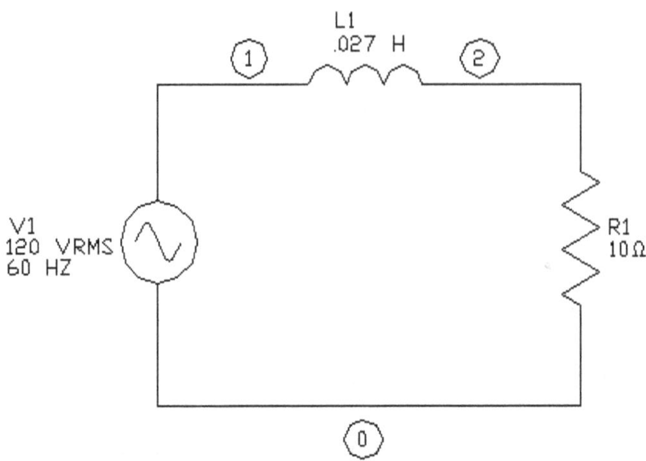

Figure 3-2-1 AC voltage applied to a resistor in series with an inductor.

Description	Symbol	Value and units
Fundamental AC voltage	V1	.0166667 second period, 60 Hz, sinusoidal waveform. 120 V_{ACrms} = 169.7 V_{ACpeak} .
Inductance	L1	.027 H
Resistance	R1	10 Ω

PROCEDURE:

1) 120 V_{ACrms} equals 169.7 V_{ACpeak}. Expressing AC voltages in their peak form is necessary in PSpice when combining AC with DC sources. It is a good habit to consistently use peak voltages in PSpice calculations.

2) Start Microsoft Notepad and enter the following netlist.

```
Figure 3-2-1.CIR RESISTOR INDUCTOR DIVIDER WITH AC INPUT
* Finding voltage across R1
V1 1 0 AC 169.7 0
L1 1 2 .027
R1 2 0 10
.AC LIN 1 60 60
.PRINT AC V(2)
.END
```

3) Line by line description of the new types of statement lines in this netlist program:

Line #3, "V1 1 0 AC 169.7 0", describes the voltage supply, V1. The "1" on the left shows that the location of positive voltage is at node 1 when $0 < (2\pi ft + \text{phase angle}) < \pi$ radians. The first "0" shows the location of the corresponding negative voltage is at node 0. The "AC" shows that the "V1" voltage is AC. The "169.7" shows that the peak voltage between node 1 and node 0 is 169.7 volts. The second "0" is the voltage's phase angle in degrees.

Line #4, "L1 1 2 .027", describes the inductor between node 1 and node 2. The ".027" is the inductor's inductance in H.

Line #6, ".AC LIN 1 60 60" describes the range of frequency of the AC source voltage. The "LIN" means that frequencies are to be varied linearly from the lowest frequency to the highest. The "1" means that there is only one frequency step selected. The first "60" is the low limit of frequency. The second "60" is the high limit of frequency. This statement could be used to sweep the frequency over a range of frequencies by changing the second "60" and putting a larger integer number than 1 for the number of frequencies.

4) Run the PSpice simulation and look at the output file.

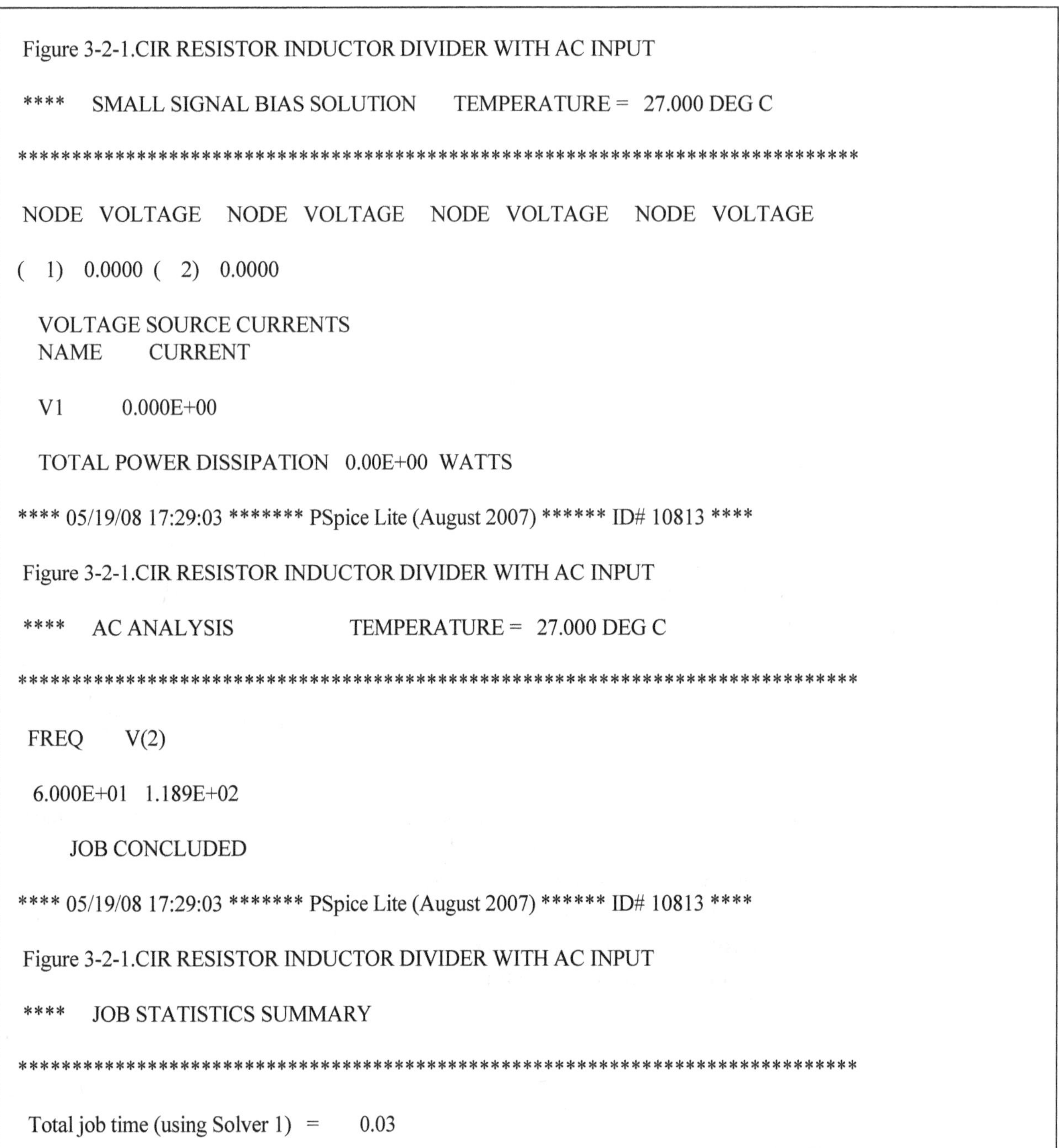

```
Figure 3-2-1.CIR RESISTOR INDUCTOR DIVIDER WITH AC INPUT

****    SMALL SIGNAL BIAS SOLUTION      TEMPERATURE =  27.000 DEG C

******************************************************************************

NODE  VOLTAGE    NODE  VOLTAGE    NODE  VOLTAGE    NODE  VOLTAGE

(  1)  0.0000 (  2)  0.0000

  VOLTAGE SOURCE CURRENTS
  NAME        CURRENT

  V1          0.000E+00

  TOTAL POWER DISSIPATION   0.00E+00  WATTS

**** 05/19/08 17:29:03 ******* PSpice Lite (August 2007) ****** ID# 10813 ****

Figure 3-2-1.CIR RESISTOR INDUCTOR DIVIDER WITH AC INPUT

****    AC ANALYSIS                   TEMPERATURE =  27.000 DEG C

******************************************************************************

 FREQ        V(2)

  6.000E+01   1.189E+02

        JOB CONCLUDED

**** 05/19/08 17:29:03 ******* PSpice Lite (August 2007) ****** ID# 10813 ****

Figure 3-2-1.CIR RESISTOR INDUCTOR DIVIDER WITH AC INPUT

****    JOB STATISTICS SUMMARY

******************************************************************************

 Total job time (using Solver 1)   =        0.03
```

Figure 3-2-2 PSpice output file from running the netlist of Figure 3-2-1.

5) In the "SMALL SIGNAL BIAS SOLUTION" section of the output, note that the DC voltages are all 0.

6) In the "AC ANALYSIS" section of the output, note that voltage from node 2 to node 0, V(2), is 118.9 V_{ACpeak}.

3.3 TRANSIENT ANALYSIS

3.3.1 TRANSIENT ANALYSIS OF A STEADY-STATE AC CIRCUIT

PROBLEM: The problem of Section 3.2 is solved again, but this time using PSpice's transient analysis capability. Here plots will be made of the time dependent source voltage and output voltage versus time with the PSpice Probe program.

Description	Symbol	Value and units
Fundamental AC sine wave voltage	V1	0 V offset, 169.7 V_{ACpeak}, 60 Hz frequency, 0 second delay time, 0 sec^{-1} damping factor, and 0° phase angle.
Inductance	L1	.027 H
Resistance	R1	10 Ω

PROCEDURE:

1) Start Microsoft Notepad and enter the following netlist.

```
Figure 3-2-1transient.CIR RESISTOR INDUCTOR DIVIDER WITH AC INPUT
* Plotting voltage across V1 and R1 versus time with .PROBE
V1 1  0 SIN(0 169.7 60 0 0 0)
L1 1 2 .027
R1 2 0 10
.TRAN .1m 40m
.PROBE
.END
```

2) Line by line description of the new statement lines in this netlist program:

Line #3, "V1 1 0 SIN(0 169.7 60 0 0 0)", describes the sinusoidal voltage supply. The "1" on the left shows that the location of positive voltage is at node 1 when $0 < (2\pi ft + \text{phase angle}) < \pi$ radians. The "0" shows the location of the corresponding negative voltage is at node 0. The "SIN" shows that the "V1" voltage is sinusoidal AC. Inside the parentheses starting on the left: The "0" shows that there is 0 DC volts offset from the average voltage. The "169.7" shows that the sine peak voltage between node 1 and node 0 is 169.7 volts. The "60" is the frequency of the sine wave in Hz. The next "0" is the delay time before the sine wave starts in seconds. The next "0" is the damping factor. That "0" means the sine wave is undamped. The last "0" is the phase angle of the sine wave in degrees.

Line #6, ".TRAN .1m 40m", shows that this is a transient analysis. The ".1m" indicates that a time step of .1 millisecond would be used if the output is displayed with ".PLOT" or ".PRINT" outputs. It is not necessarily the computing step time. PSpice usually selects computing step times much smaller than that requested in the .TRAN statement. The "40m" is the 40 millisecond time span of the analysis.

Line #7, ".PROBE" starts the graphical output display program. Probe displays PSpice voltages, currents, and powers versus time or frequency. Its very useful oscilloscope-like display will be used in the book's transient analysis examples.

3) Run the PSpice simulation. A blank Probe graph will automatically appear on screen. The time base of the graph will go from 0 to 40 milliseconds.

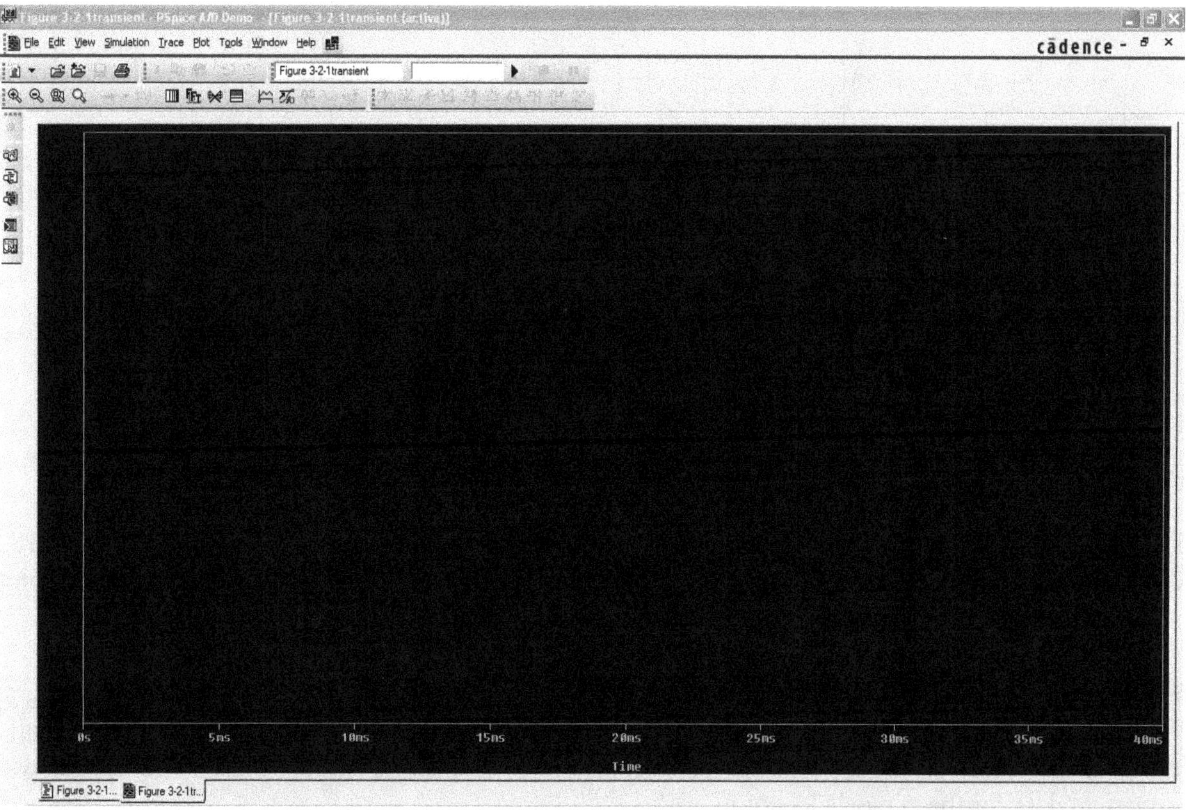

Figure 3-3-1-1 Blank Probe output graph from running a transient PSpice analysis on the circuit of Figure 3-2-1.

4) In the main tool bar select "Trace" and under that select "Add Traces". Under "Add Traces" either select V(1) and V(2) from the list or simply type in V(1) and V(2). V(1) is the voltage between node 1 and node 0. V(1) could also be written V(1,0). In the same way, V(2) is the voltage between node 2 and node 0 and could be written V(2,0).The graph in Figure 3-3-1-2 will be produced.

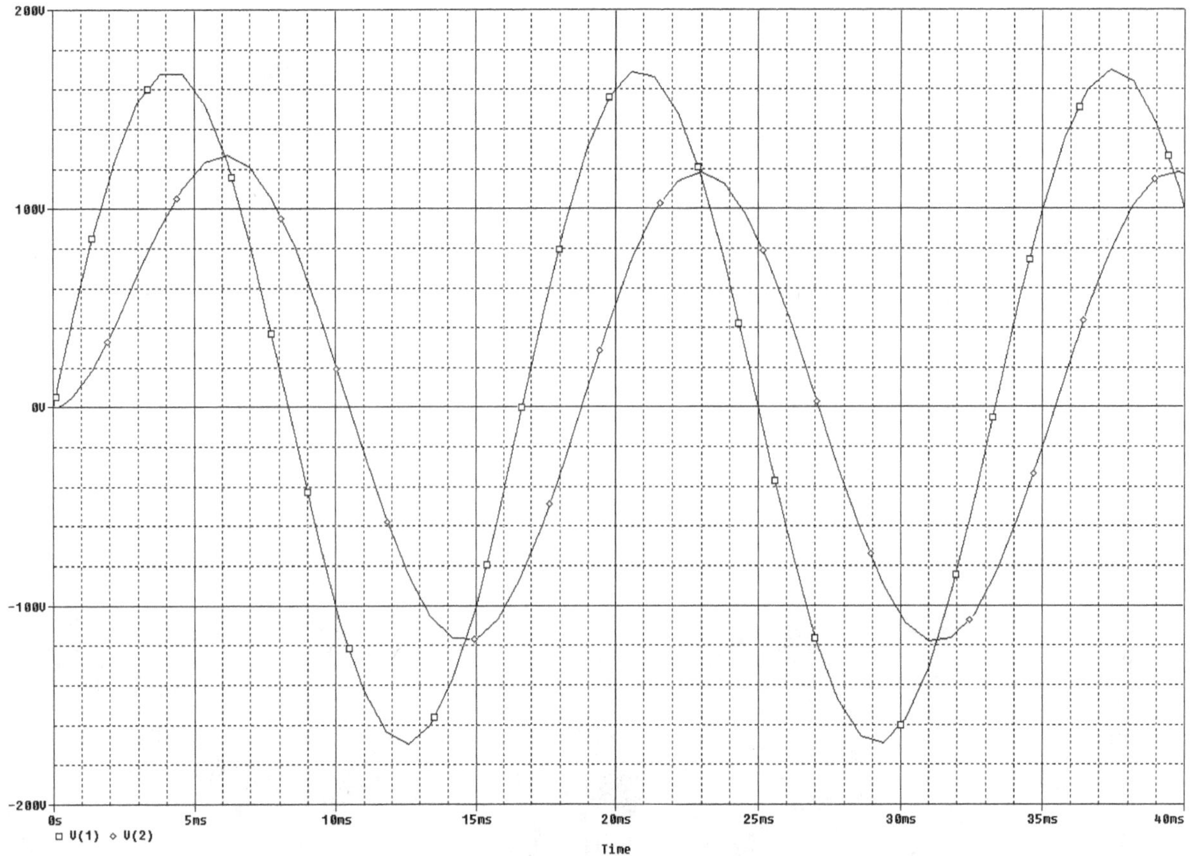

Figure 3-3-1-2 Probe output graph from the transient PSpice analysis of the circuit of Figure 3-2-1.

5) In the Probe output graph, after V(2) has decreased to its steady-state value, V(2) is 118.9 V_{ACpeak}. Note the jaggedness of the plotted waveforms. The Probe program chooses to plot 42 points on each curve and to fill in the spaces between the points with straight-line interpolations.

6) In the Probe program numerical values of the waveforms can be determined by left clicking on "Trace", "Cursor", and then the symbol of the waveform at the bottom left of the display. Finally, left click the cursor on the appropriate location on the waveform. A small window will appear with numerical data for the cursor's location.

7) Older methods for outputting PSpice data used the .PLOT and .PRINT statements. .PLOT will produce a numerical value and plot point for variables for each time step in the .TRAN statement. The .PRINT will simply print numerical values for variables for each time step in the .TRAN statement.

8) The following netlist includes a .PLOT statement. The .PLOT statement creates a data chart and plot for V(1) and V(2) in the output file.

```
Figure 3-2-1transientplot.CIR RESISTOR INDUCTOR DIVIDER WITH AC INPUT
* Plotting voltage across V1 and R1 versus time with .PLOT
V1  1  0  SIN(0  169.7  60  0  0  0)
L1  1  2  .027
R1  2  0  10
.TRAN  .1m  40m
.PLOT  TRAN  V(1)  V(2)
.PROBE
.END
```

9) When this is run the Probe display is the same as before. However the output file now contains data and a plot for V(1) and V(2). Part of this is shown in Figure 3-3-1-3.

```
**** 05/19/08 17:58:33 ******* PSpice Lite (August 2007) ****** ID# 10813 ****

 Figure 3-2-1transientplot.CIR RESISTOR INDUCTOR DIVIDER WITH AC INPUT

 ****     TRANSIENT ANALYSIS              TEMPERATURE =   27.000 DEG C

******************************************************************************

 LEGEND:

*: V(1)
+: V(2)

  TIME           V(1)
(*+)-------------      -2.0E+2  -1.0E+2  0.0E+0  1.0E+2  2.0E+2
                        ------------------------------
 0.000E+00  0.000E+00 .        .        X        .        .
 1.000E-04  6.396E+00 .        .        +*       .        .
 2.000E-04  1.278E+01 .        .        + *      .        .
 3.000E-04  1.914E+01 .        .        + *      .        .
 4.000E-04  2.546E+01 .        .        +  *     .        .
 5.000E-04  3.173E+01 .        .        +   *    .        .
 6.000E-04  3.800E+01 .        .       .+   *    .        .
 7.000E-04  4.426E+01 .        .       .+    *   .        .
 8.000E-04  5.015E+01 .        .       .+     *  .        .
 9.000E-04  5.604E+01 .        .       .+     *  .        .
 1.000E-03  6.194E+01 .        .       .+      * .        .
 1.100E-03  6.783E+01 .        .       . +     * .        .
 1.200E-03  7.373E+01 .        .       . +      *.        ..
.
.
```

Continued

```
.
.
 3.910E-02  1.390E+02 .         .        .        .+  *        .
 3.920E-02  1.349E+02 .         .        .        .+  *        .
 3.930E-02  1.307E+02 .         .        .        .+ *         .
 3.940E-02  1.265E+02 .         .        .        .+*          .
 3.950E-02  1.224E+02 .         .        .        .+*          .
 3.960E-02  1.182E+02 .         .        .        .X           .
 3.970E-02  1.141E+02 .         .        .        .X           .
 3.980E-02  1.098E+02 .         .        .        .*+          .
 3.990E-02  1.048E+02 .         .        .        .*+          .
 4.000E-02  9.975E+01 .         .        .        * +          .
                      - - - - - - - - - - - - - - - - - - - - - - - - - - -
```

Figure 3-3-1-3 Part of the .PLOT output file for the transient PSpice analysis of the circuit of Figure 3-2-1. Note the horizontal axis values of V(1) have been manually truncated to allow them to better fit on the page.

3.3.2 TRANSIENT ANALYSIS OF A SQUARE-WAVE CIRCUIT

PROBLEM: A 120 V_{ACrms} square-wave supply is connected in series with the same .027 H inductor and 10 Ω resistor of Section 3.2. Using the PSpice and Probe, what is the voltage waveform across the 10 Ω resistor?

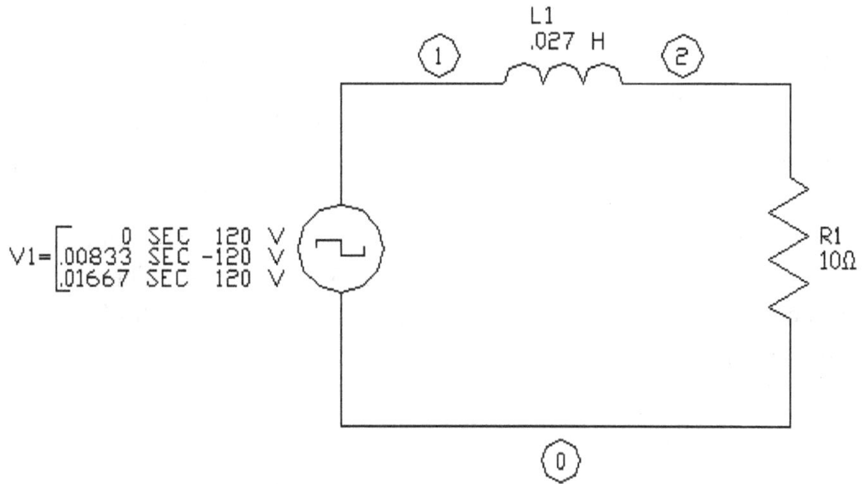

Figure 3-3-2-1 Pulse source supplying a square-wave to a resistor in series with an inductor.

Description	Symbol	Value and units
Fundamental AC pulse voltage	V1	-120 V initial voltage, 120 V pulsed voltage, 0 second time delay, .2 millisecond rise time, .2 millisecond fall time, .00833333 second pulse width, .0166667 second period
Inductance	L1	.027 H
Resistance	R1	10 Ω

PROCEDURE:

1) Start Microsoft Notepad and enter the following netlist.

```
Figure 3-3-2-1.CIR RESISTOR INDUCTOR DIVIDER WITH SQUARE-WAVE INPUT
* Plotting voltage across V1 and R1 versus time with .PROBE
V1  1  0  PULSE(-120V  120V  0  .2m  .2m  .00833333  .0166667)
L1  1  2  .027
R1  2  0  10
.TRAN  .1m  40m
.PROBE
.END
```

2) Description of the new statement line in this netlist program:

Line #3, "V1 1 0 PULSE(-120V 120V 0 .2m .2m .00833333 .0166667)", describes the square-wave voltage supply. The "1" on the left shows that the location of the pulse voltage is at node 1 when the pulse is on. The "0" shows the location of the other polarity is at node 0. The "PULSE" shows that the "V1" voltage is the pulse voltage. The square-wave is made of repeating positive pulses followed by negative base voltages. Inside the parentheses starting on the left: The "-120V" is the base voltage when the pulse first starts. The "120V" is the voltage that the pulse rises to. The first "0" is the 0 second time delay before the voltage first starts to rise from "-120V". The first ".2m" is the .2 milliseconds the voltage takes to rise from –120V to 120V. The second ".2m" is the .2 milliseconds the voltage takes to fall from 120V to –120V. The ".00833333" is the number of seconds that the pulse remains at 120V. The ".0166667" is total number of seconds before the pulse pattern repeats.

3) Run the PSpice simulation and Probe.

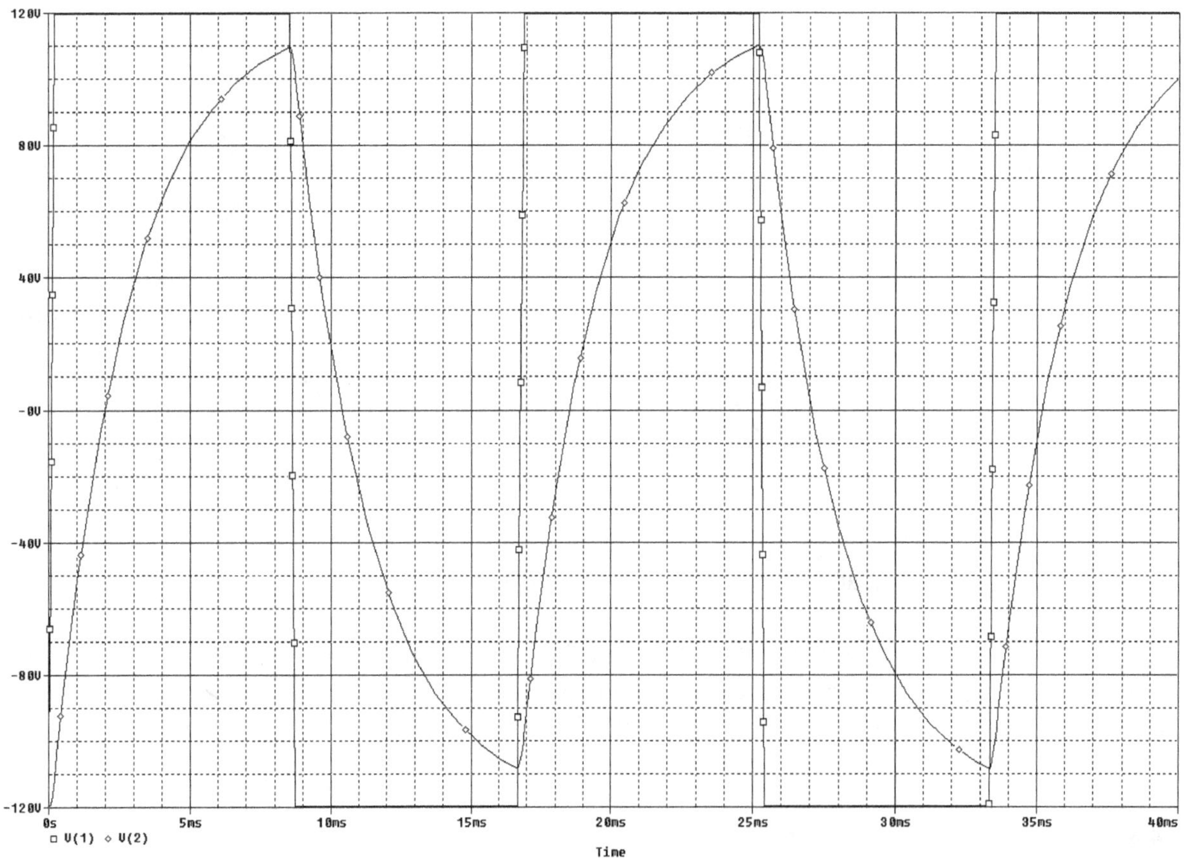

Figure 3-3-2-2 Probe output graph from running a transient PSpice analysis on the circuit of Figure 3-3-2-1.

4.0 POWER ELECTRONIC EXAMPLE PROBLEMS

Power electronic control devices such as SCRs (**S**ilicon-**C**ontrolled **R**ectifiers), GTOs **G**ate **T**urn-**O**ff Thyristors), IGBTs (**I**nsulated-**G**ate **B**ipolar **T**ransistors) and IGCTs (**I**ntegrated-**G**ate **C**ommutated **T**hyristors) are basically on/off devices that can be modeled by on/off switches and/or diodes.

In each of the following simulations the power electronic circuit is drawn. Then the circuit is redrawn with on/off switches and/or diodes in place of the power electronic control devices. Additional components are added to the circuits as needed.

Note that the computing time for some of these simulations can be several minutes.

4.1 AC TO DC CONVERTER

PROBLEM: In a mine, a three-phase 575 V_{ACrms} line-to-neutral 60 Hz supply is connected through a 2500 ft. transmission line to an SCR converter that supplies variable DC voltage to one 85 HP 1344 V_{DC} series motor. The three-phase voltage at the supply has repeating notches in it as seen in the line-to-neutral voltage waveform of Figure 4-1-1. It is physically difficult to measure the voltages at the motor and converter because the terminals are enclosed in explosion-proof housings. PSpice is used to predict the voltages at the converter input, across the converter's SCRs, and across the DC motor input.

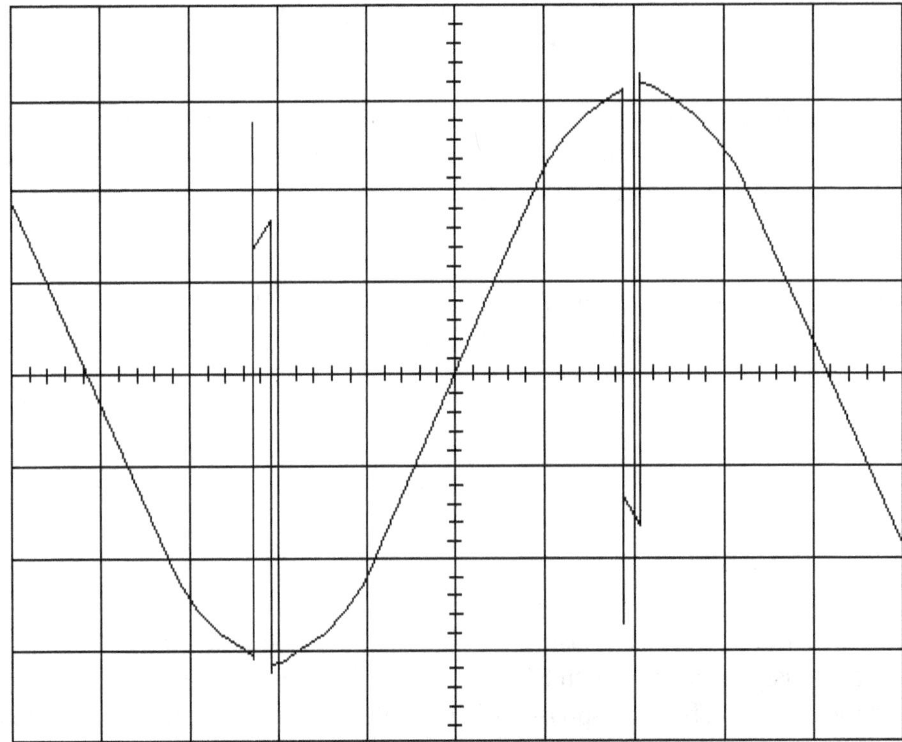

Figure 4-1-1 Oscillograph of one phase of the line-to-neutral supply voltage.

4.1.1 AC TO DC CONVERTER PRODUCING FULL-LOAD

PROCEDURE:

1) Draw the schematic.

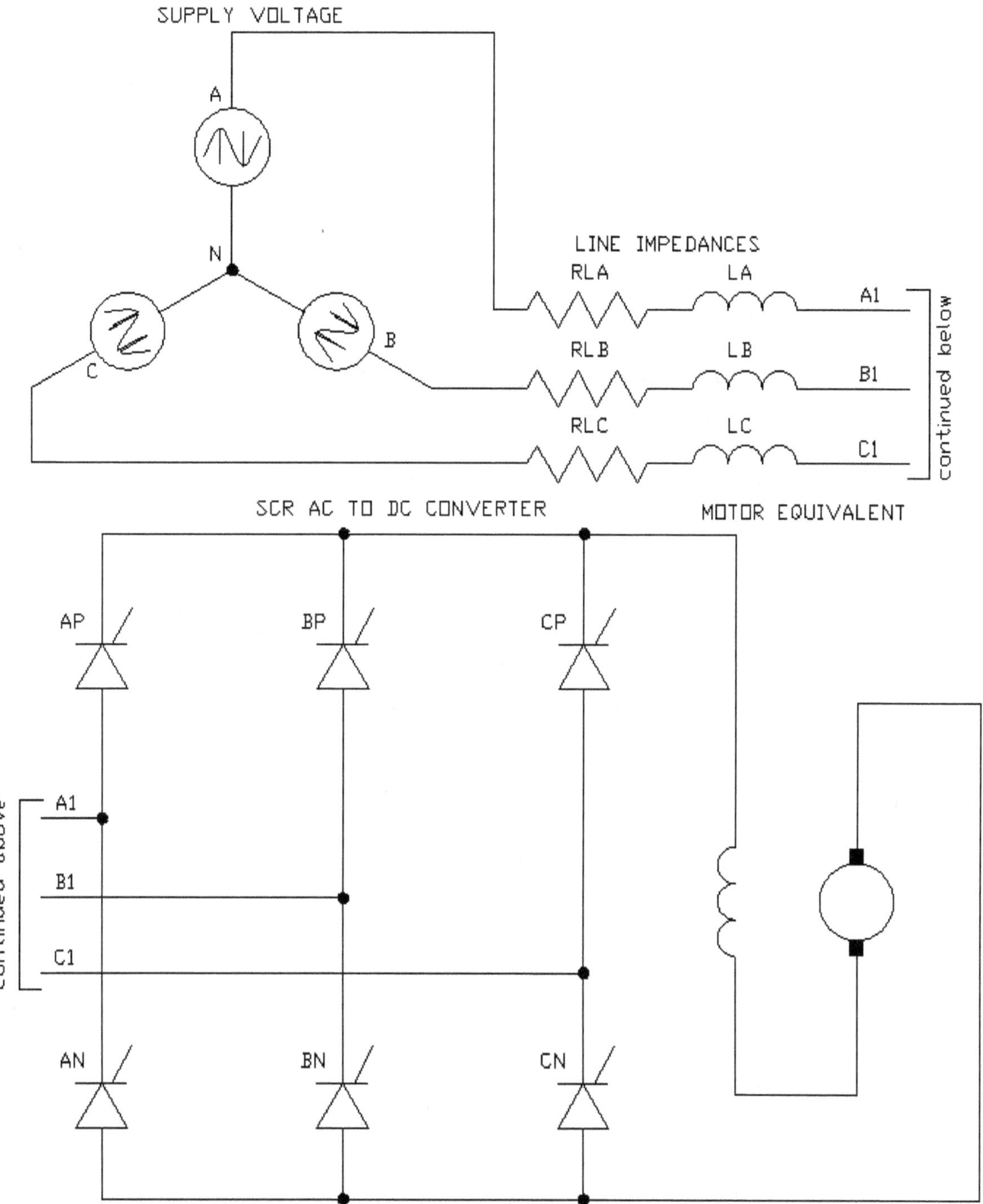

Figure 4-1-1-1 575 V_{ACrms} 60 Hz supply to an SCR converter supplying an 85 HP DC series motor.

2) Select values for the variables.

The line will be assumed to be 1/0 AWG cable laying loosely in a cable tray. Calculations show the resistance per phase and inductance per phase should be about .25 Ω and .0004 H. From this: RLA = RLB = RLC = .25 Ω and LLA = LLB = LLC = .0004 H.

An effective resistance and inductance will be inserted in the circuit in place of the steady-state DC series motor. The DC converter will be assumed to be producing a maximum voltage. This is the voltage that would be produced when the SCRs are always on. When the SCRs are always on, they can be considered equivalent to diodes. The effective DC output voltage of the converter is then 1.35 x SQR(3) x 575 = 1344 V_{DC}.

The DC motor will be assumed to be operating at a full-load of 85 HP at 85% efficiency. The effective motor resistance, RM, is calculated as $(1344)^2$/([85][746]/.85) = 24.2 Ω.

The effective motor inductance, LM, is estimated to be .0457 H.

3) Redraw the circuit with numbered nodes, SCRs drawn as diodes, and the motor drawn as an equivalent resistance and inductance. Diodes are used in place of SCRs to simplify the full-load analysis. SCRs in a converter producing maximum output are essentially diodes. Later, in Section 4.1.2, analysis with controlled switches and diodes in place of the SCRs will be done to determine operation characteristics at lesser loads.

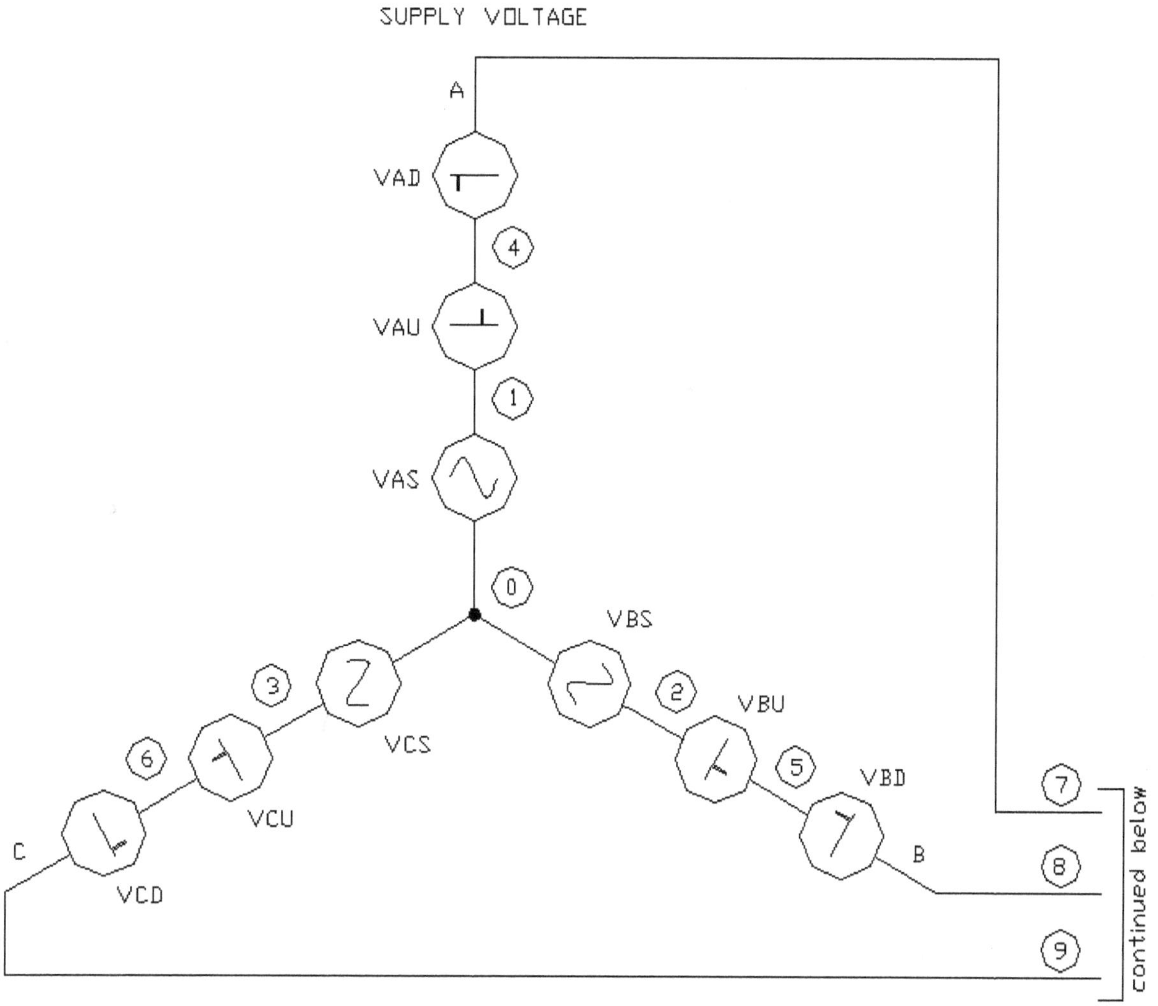

SUPPLY VOLTAGE
A
VAD
4
VAU
1
VAS
0
VBS
2
VBU
5
VBD
B
VCS
3
VCU
6
VCD
C
7
8
9
continued below

LINE IMPEDANCES
continued above
7
RLA
10
LLA
13
8
RLB
11
LLB
14
9
RLC
12
LLC
15
continued on following page

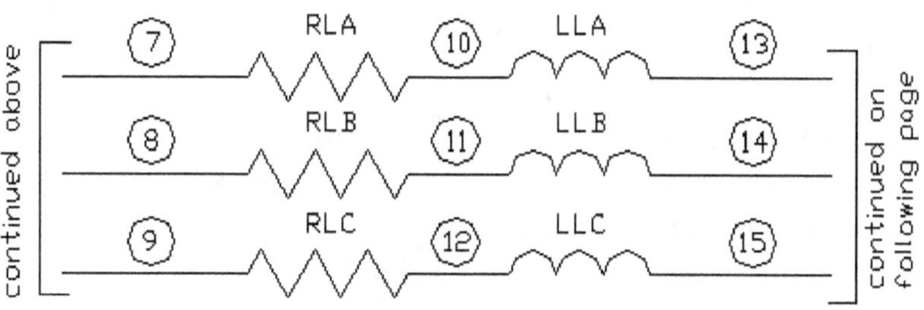

Continued

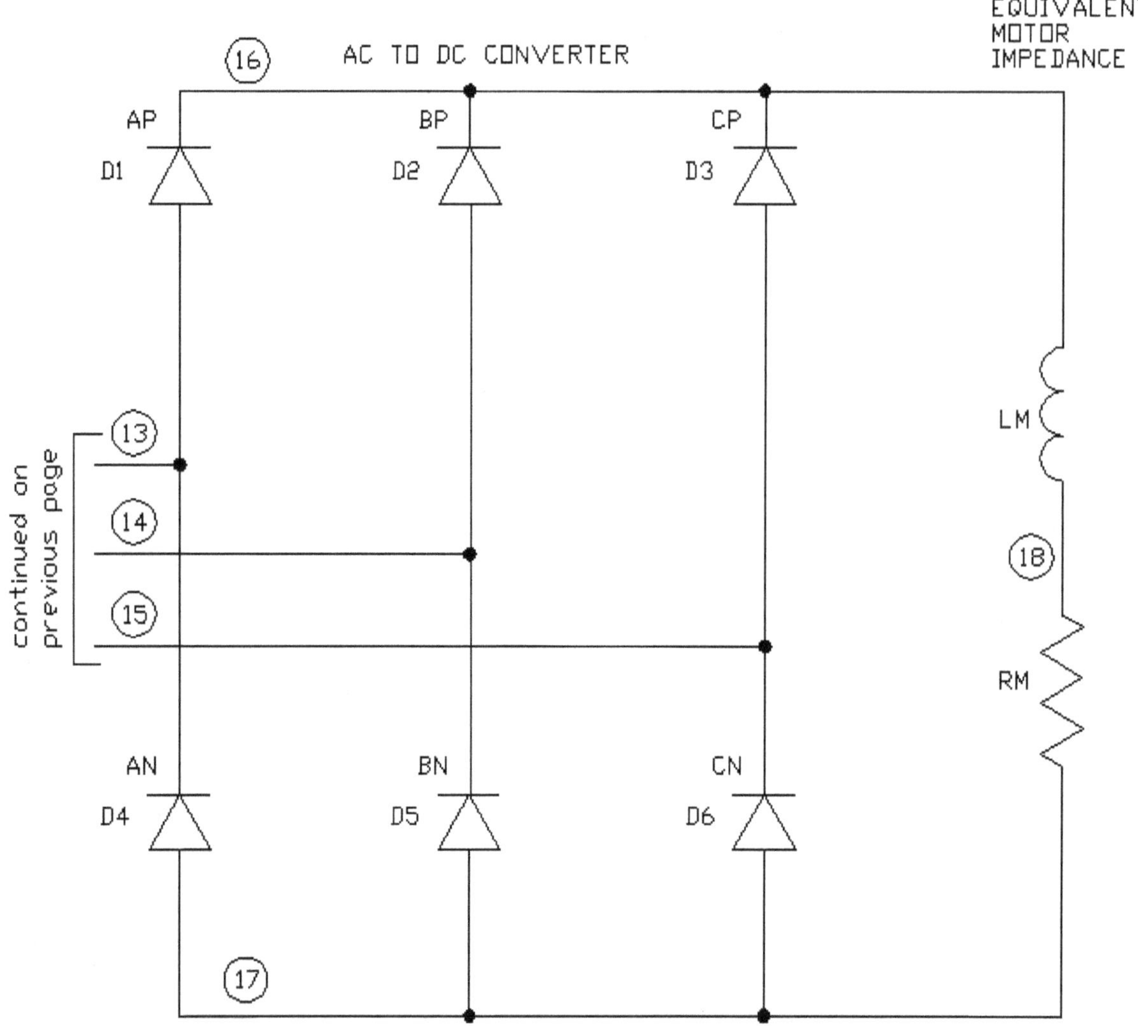

Figure 4-1-1-2 Nodes numbered for use in the netlist, the SCRs have been changed to diodes, and the DC motor load changed to an equivalent resistance and inductance.

4) Specify the voltage supply waveforms. The oscillograph of Figure 4-1-1 is used to create three waveforms, one for each of the phases. Then each of the phase waveforms is separated into a sinusoidal voltage, a negative pulse, and a positive pulse. PSpice repeats each sinusoidal voltage and pulse every period. The waveforms are shown in Figure 4-1-1-3.

5) Values for the supply voltage and components are:

Description	Symbol	Value and units
Fundamental AC sinusoidal voltage	VAS	0 V offset, 813.2 V_{ACpeak}, 60 Hz frequency, 0 second delay time, 0 sec^{-1} damping factor, and 0° phase angle. This is shown in Figure 4-1-1-3.
Fundamental AC sinusoidal voltage	VBS	0 V offset, 813.2 V_{ACpeak}, 60 Hz frequency, 0 second delay time, 0 sec^{-1} damping factor, and -120° phase angle. This is shown in Figure 4-1-1-3.
Fundamental AC sinusoidal voltage	VCS	0 V offset, 813.2 V_{ACpeak}, 60 Hz frequency, 0 second delay time, 0 sec^{-1} damping factor, and -240° phase angle. This is shown in Figure 4-1-1-3.
Positive notch voltage	VAU	0 V initial voltage, 1125 V pulsed voltage, .01202 second time delay, 0 second rise time, 0 second fall time, .0004 second pulse width, .0166667 second period. This is shown in Figure 4-1-1-3.
Positive notch voltage	VBU	0 V initial voltage, 1125 V pulsed voltage, .0009089 second time delay, 0 second rise time, 0 second fall time, .0004 second pulse width, .0166667 second period. This is shown in Figure 4-1-1-3.
Positive notch voltage	VCU	0 V initial voltage, 1125 V pulsed voltage, .00646 second time delay, 0 second rise time, 0 second fall time, .0004 second pulse width, .0166667 second period. This is shown in Figure 4-1-1-3.

Continued

Negative notch voltage	VAD	0 V initial voltage, -1125 V pulsed voltage, .00369 second time delay, 0 second rise time, 0 second fall time, .0004 second pulse width, .0166667 second period. This is shown in Figure 4-1-1-3.
Negative notch voltage	VBD	0 V initial voltage, -1125 V pulsed voltage, .00925 second time delay, 0 second rise time, 0 second fall time, .0004 second pulse width, .0166667 second period. This is shown in Figure 4-1-1-3.
Negative notch voltage	VCD	0 V initial voltage, -1125 V pulsed voltage, .01480 second time delay, 0 second rise time, 0 second fall time, .0004 second pulse width, .0166667 second period. This is shown in Figure 4-1-1-3.
Line resistance	RLA, RLB, RLC	.25 Ω
Line inductance	LLA, LLB, LLC	.0004 H
Diodes	D1 to D6	PLAIN diodes using the PSpice default parameters for simple ideal diodes
Equivalent motor inductance	LM	.0457 H
Equivalent motor resistance	RM	24.2 Ω

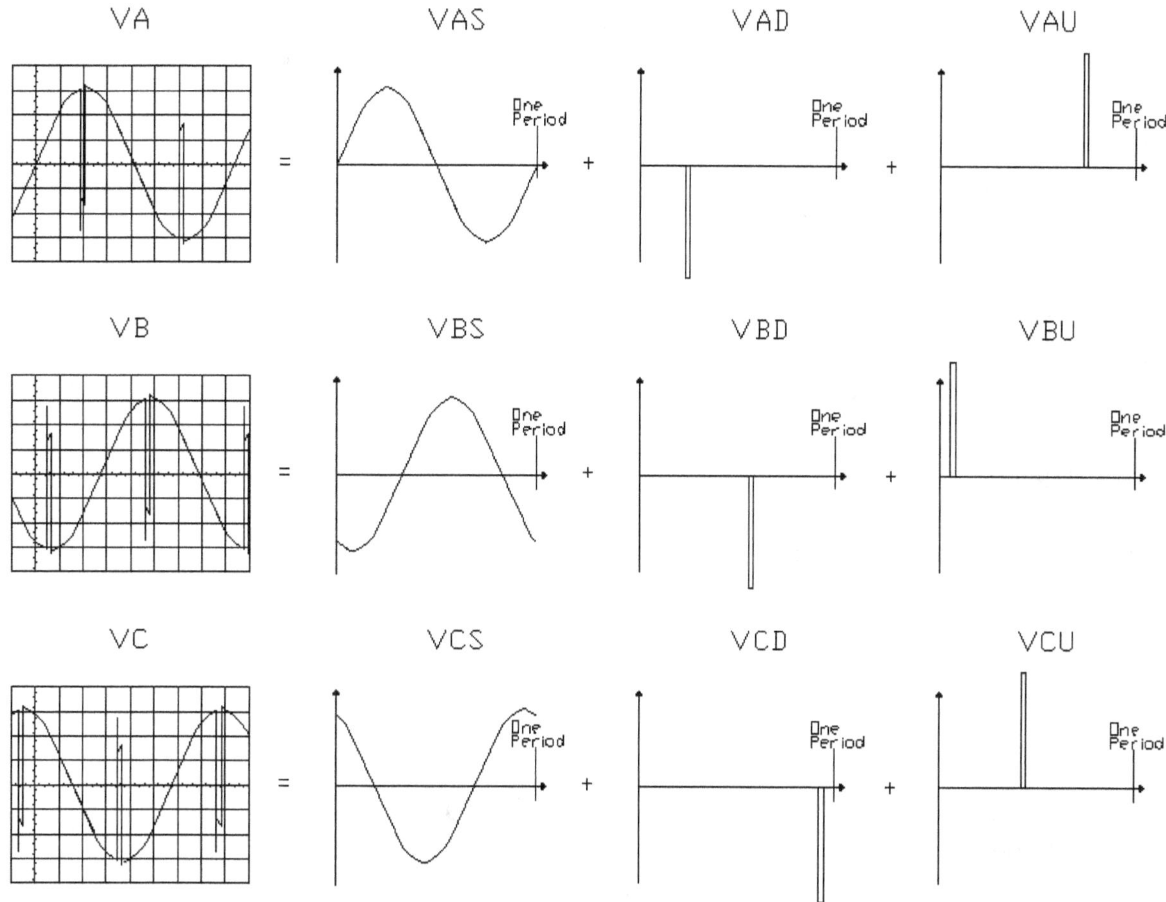

Figure 4-1-1-3 Phase voltages decomposed into sinusoid and pulse voltages.

6) The netlist program is:

```
Figure 4-1-1-2.CIR CONVERTER AT FULL-LOAD WITH NOTCHED INPUT VOLTAGES
* Plotting source, converter, and load voltages versus time and finding the
* output form factor with continuously turned on SCRs
VAS 1 0 SIN(0 813.2 60 0 0 0)
VBS 2 0 SIN(0 813.2 60 0 0 -120)
VCS 3 0 SIN(0 813.2 60 0 0 -240)
*
VAU 4 1 PULSE(0 1125V .01202 0 0 .0004 .0166667)
VBU 5 2 PULSE(0 1125V .0009089 0 0 .0004 .0166667)
VCU 6 3 PULSE(0 1125V .00646 0 0 .0004 .0166667)
*
VAD 4 7 PULSE(0 1125V .00369 0 0 .0004 .0166667)
VBD 5 8 PULSE(0 1125V .00925 0 0 .0004 .0166667)
VCD 6 9 PULSE(0 1125V .0148 0 0 .0004 .0166667)
*
RLA 7 10 .25
RLB 8 11 .25
RLC 9 12 .25
*
LLA 10 13 .0004
LLB 11 14 .0004
LLC 12 15 .0004
*
D1 13 16 PLAIN
D2 14 16 PLAIN
D3 15 16 PLAIN
D4 17 13 PLAIN
D5 17 14 PLAIN
D6 17 15 PLAIN
.MODEL PLAIN D
*
LM 16 18 .0457
RM 18 17 24.2
*
.TRAN .1m 40m
*
.PROBE
.END
```

7) Description of the new statement lines in this netlist program:

The lines starting with D1, D2, D3, D4, D5, and D6 describe "PLAIN" diodes. Their left node numbers are their anode locations. Their right node numbers are their cathode locations. The "PLAIN" refers to the "PLAIN" in the line beginning with ".MODEL".

The line starting with ".MODEL" describes a simple ideal diode. "PLAIN" is the name of the diode. "D" is the description of the PSpice standard diode. Diodes other than the standard can be described by putting parameters in parentheses after the "D".

8) Once the simulation is run, Probe can be used to display traces of voltages and currents throughout the circuit.

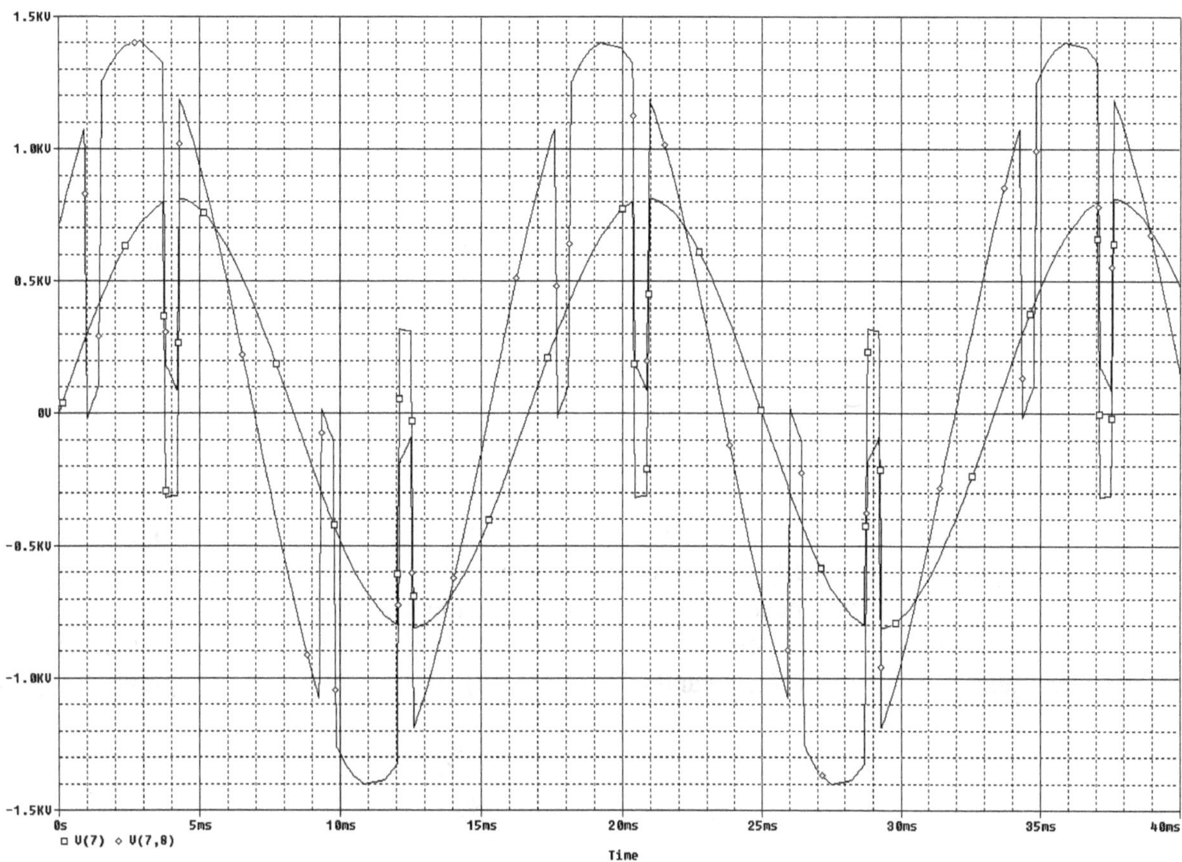

Figure 4-1-1-4 V(A-line-to-neutral) voltage (alias V(7)) and V(A-line to B-line) voltage (alias V(7,8)).

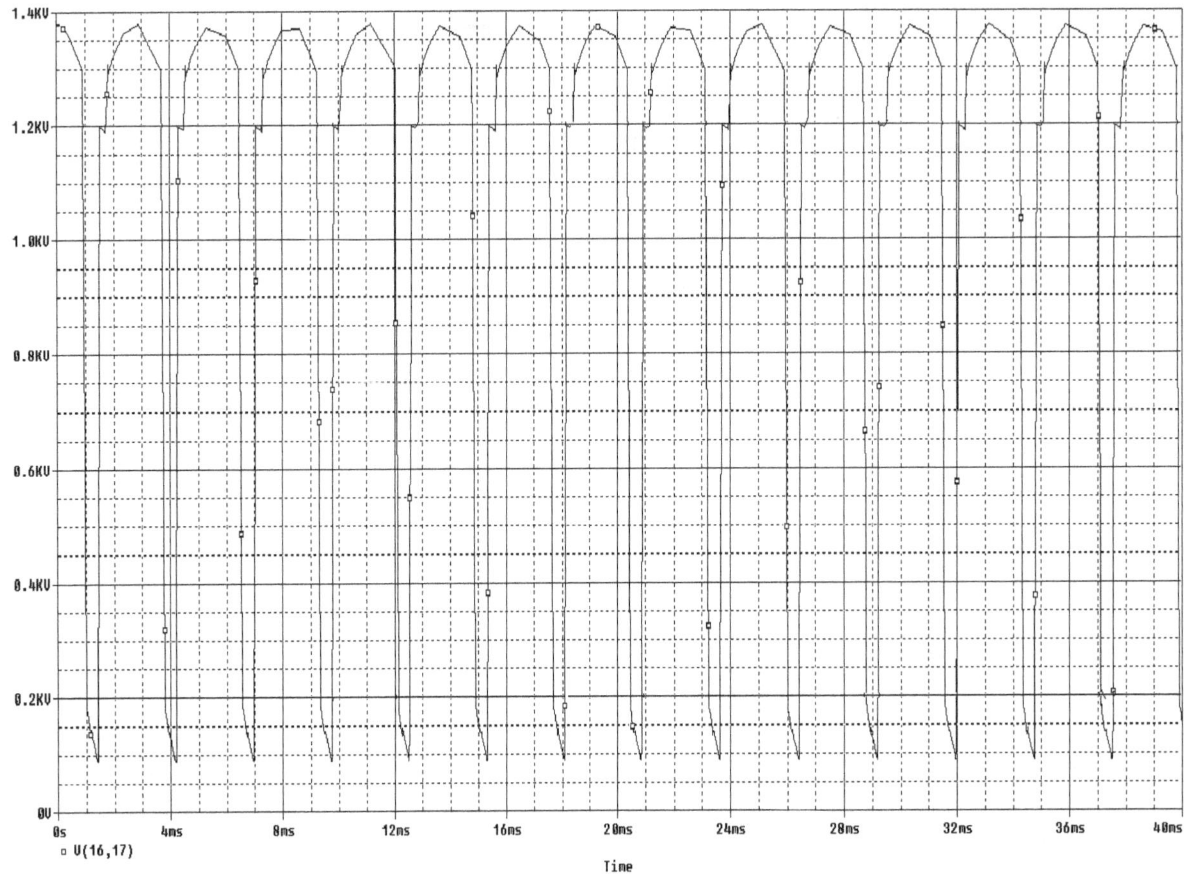

Figure 4-1-1-5 Voltage across the series motor.

9) Figure 4-1-1-5 shows that the DC being applied to the series motor has a significant AC waveform riding on it. The severity of the AC waveform rider can be determined by using Probe's AVG (average) and RMS (root-mean-square) features. With the waveform for V(16,17) of Figure 4-1-1-5 still on the computer screen, left click on "Trace" and "Delete All Traces". Now left click on "Add Traces" and enter AVG(V(16,17)) and RMS(V(16,17)). This will produce the graph of Figure 4-1-1-5.

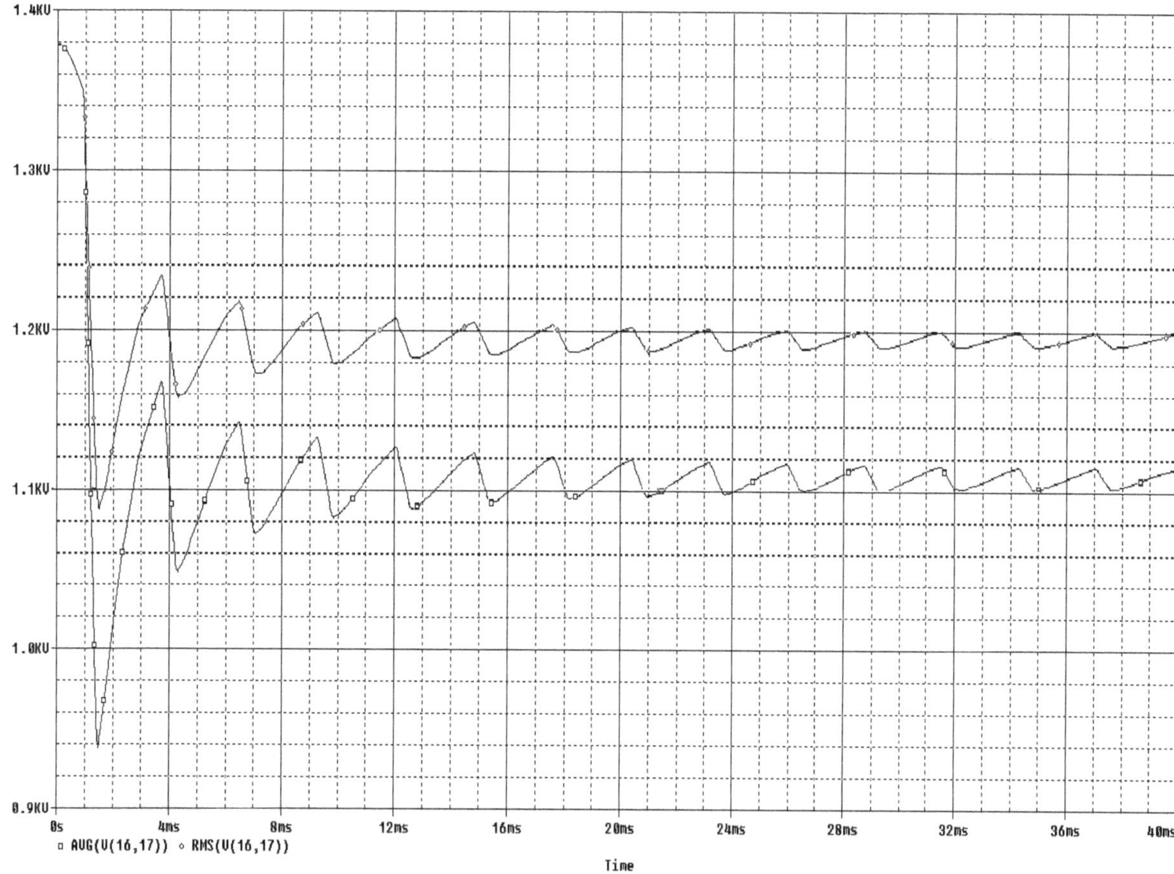

Figure 4-1-1-6 Average and rms values of the voltage across the series motor.

10) A Form Factor can be calculated for the voltage.

Form Factor = V_{ACrms}/V_{DC}

In this example at time 32 milliseconds, the Form Factor = 1.19/1.10 = 1.08

This Form Factor can be compared to that of the series motor to determine if the motor is able to withstand AC riders of this magnitude. A Form Factor of 1.08 is small; most motors should be able to withstand it.

11) Thus far, the PSpice analysis has assumed that the SCRs stay on continuously, like diodes. However, the notches on the supply voltage may be severe enough to allow the SCRs to turn off. This possibility can be checked by looking at the voltage across the diode circuit of Figure 4-1-1-2. If the voltage across the diode remains 0 continuously when the SCR should be conducting, then the notches will not turn off conducting SCRs. In Figure 4-1-1-7 that is seen to be the case, so false turn-offs are not a problem when the SCRs are turned on for whole cycles.

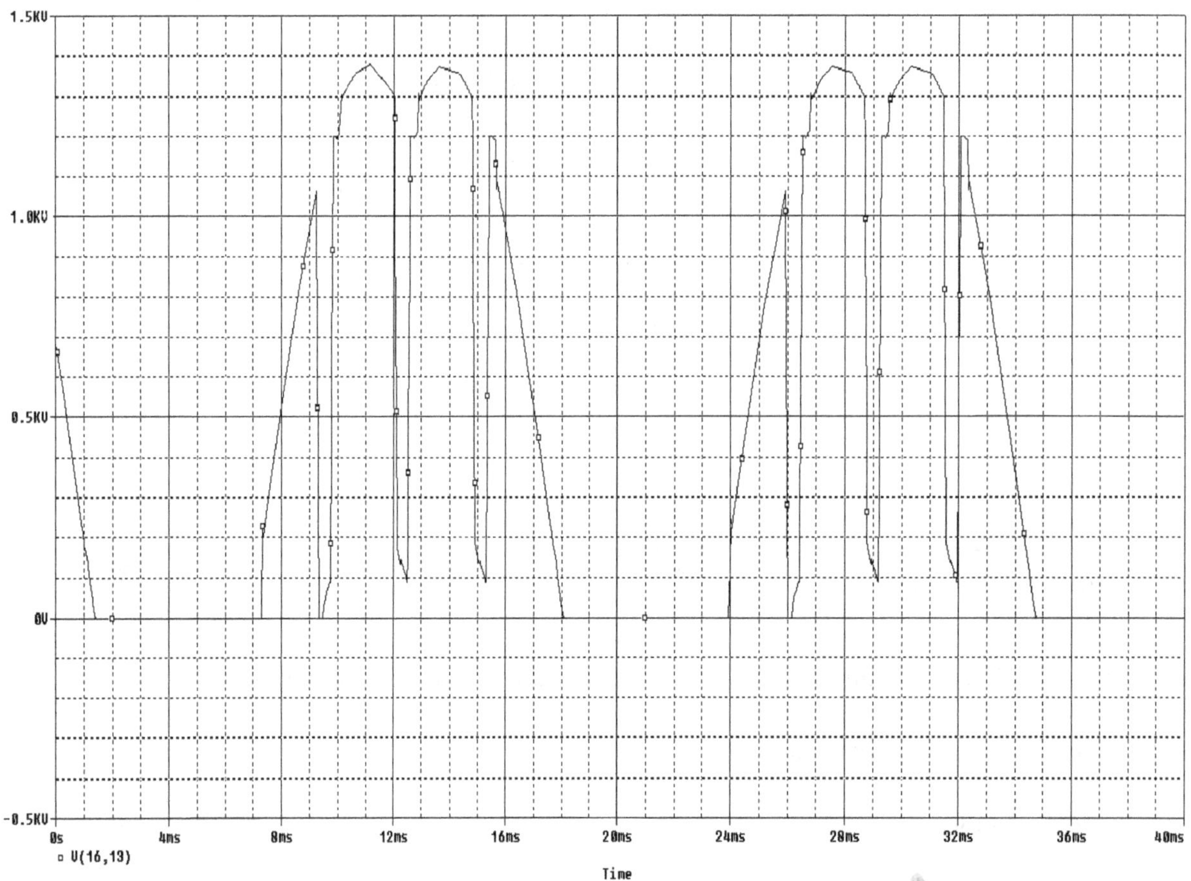

Figure 4-1-1-7 Voltage across diode D1 during full on SCR operation.

4.1.2 AC TO DC CONVERTER PRODUCING REDUCED VOLTAGE

PROCEDURE:

1) The basic schematic is the same as that shown in Figure 4-1-1-1.

2) The line-to-neutral voltage at the supply is the same as shown in Figure 4-1-1.

3) Calculate or select values for the variables.

The line resistance and inductance are the same as before.

The effective resistance of the series motor will be different. Presuming that the output power of the motor is one half of what it was at full-load, the effective resistance should be close to one half of its previous value. This means the effective resistance should be 24.2/2 = 12.1 Ω. The effective inductance should be the same as before.

4) Redraw the circuit. This time the SCRs will be on for only one half cycles, so simple diodes cannot be used. The SCRs are replaced with time controlled switches in series with diodes. In this PSpice circuit, the diodes simulate the SCRs' inability to conduct when reverse biased.

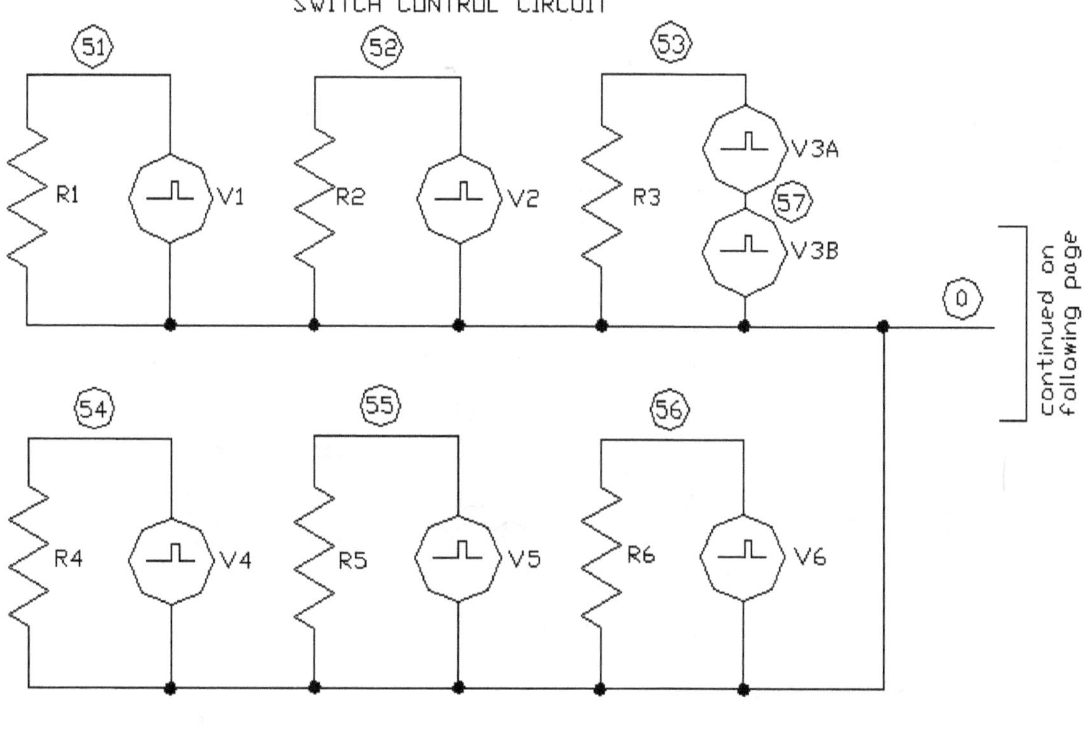

Continued

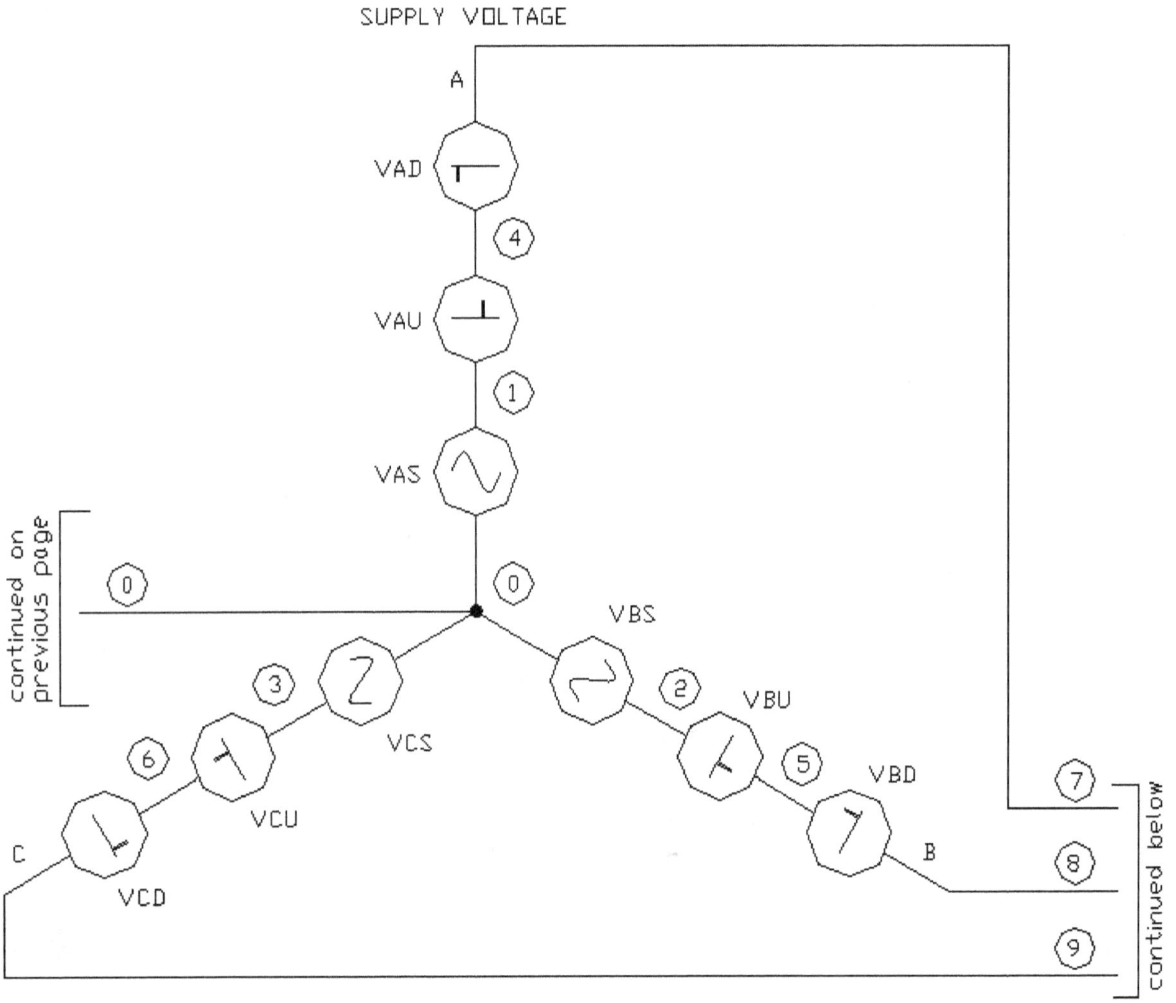

SUPPLY VOLTAGE

A
VAD
4
VAU
1
VAS
0
VBS
2
VBU
5
VBD
B
VCS
3
VCU
6
VCD
C
7
8
9

continued on previous page

continued below

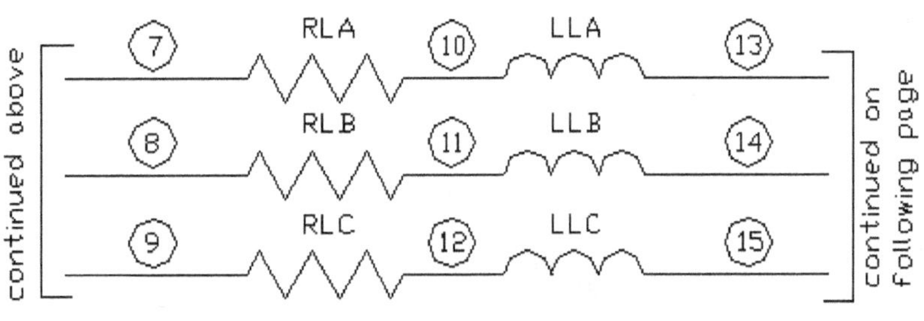

Continued

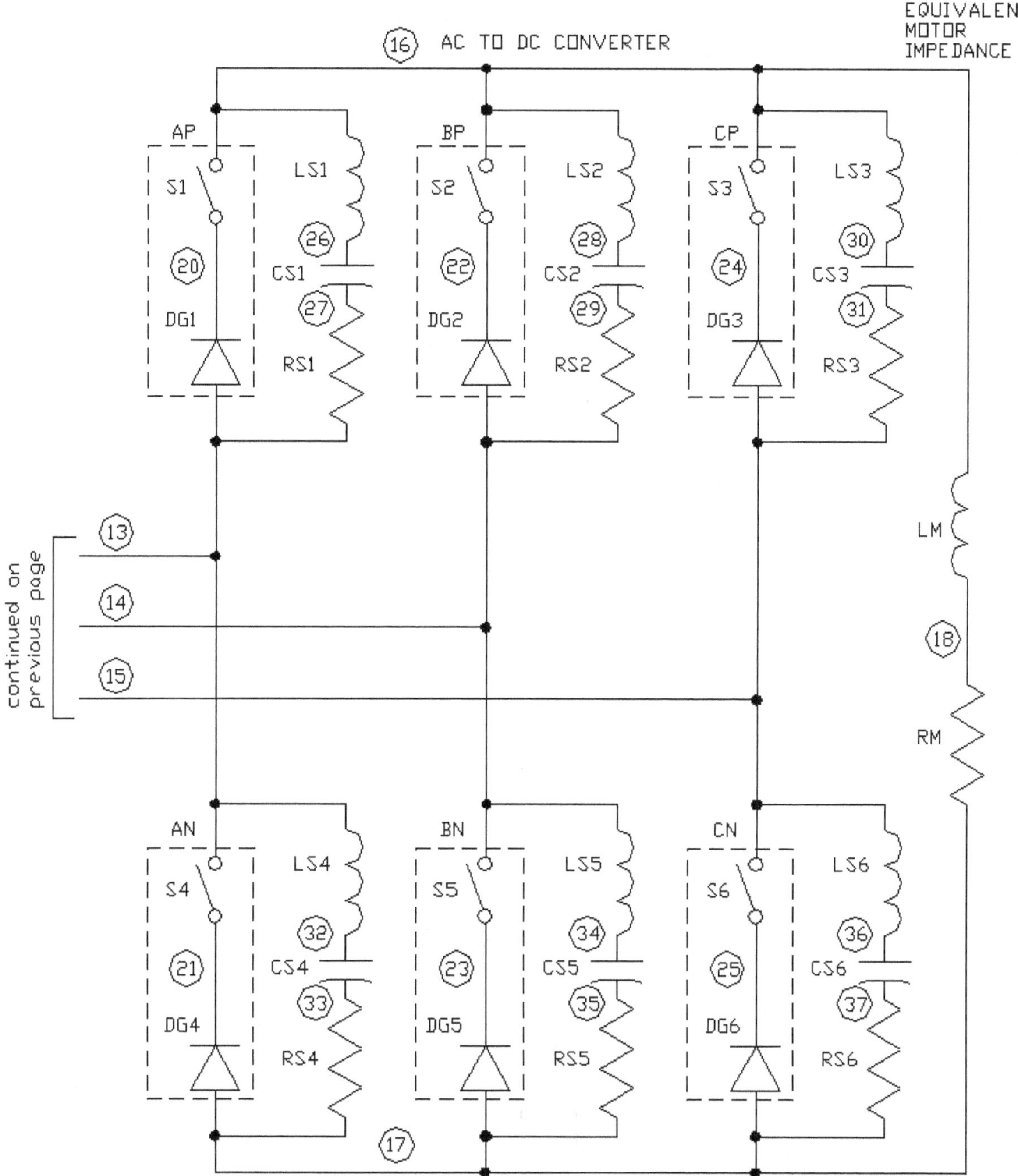

Figure 4-1-2-1 This PSpice circuit diagram is based on the circuit of Figure 4-1-1-1. The SCRs have been changed to diodes in series with timer-controlled switches, resistor/capacitor snubbers have been added, and the DC motor loads changed to an equivalent resistance and inductance.

5) Times are selected for the timer controlled switches that would conduct 90° of each SCR's cycle. Full conduction, as with the diodes in Figure 4-1-1-2, would have 120° of conduction for each cycle. The times are selected relative to the supply's three line-to-neutral voltages. Since there may be a slight voltage phase lag due to the line inductance, the selected times are not exactly at the half-voltage output point. However, they are close. If desired, the PSpice program could be run iteratively to find timer times that would come closer to producing one half output. Figure 4-1-2-2 shows the timing pulses.

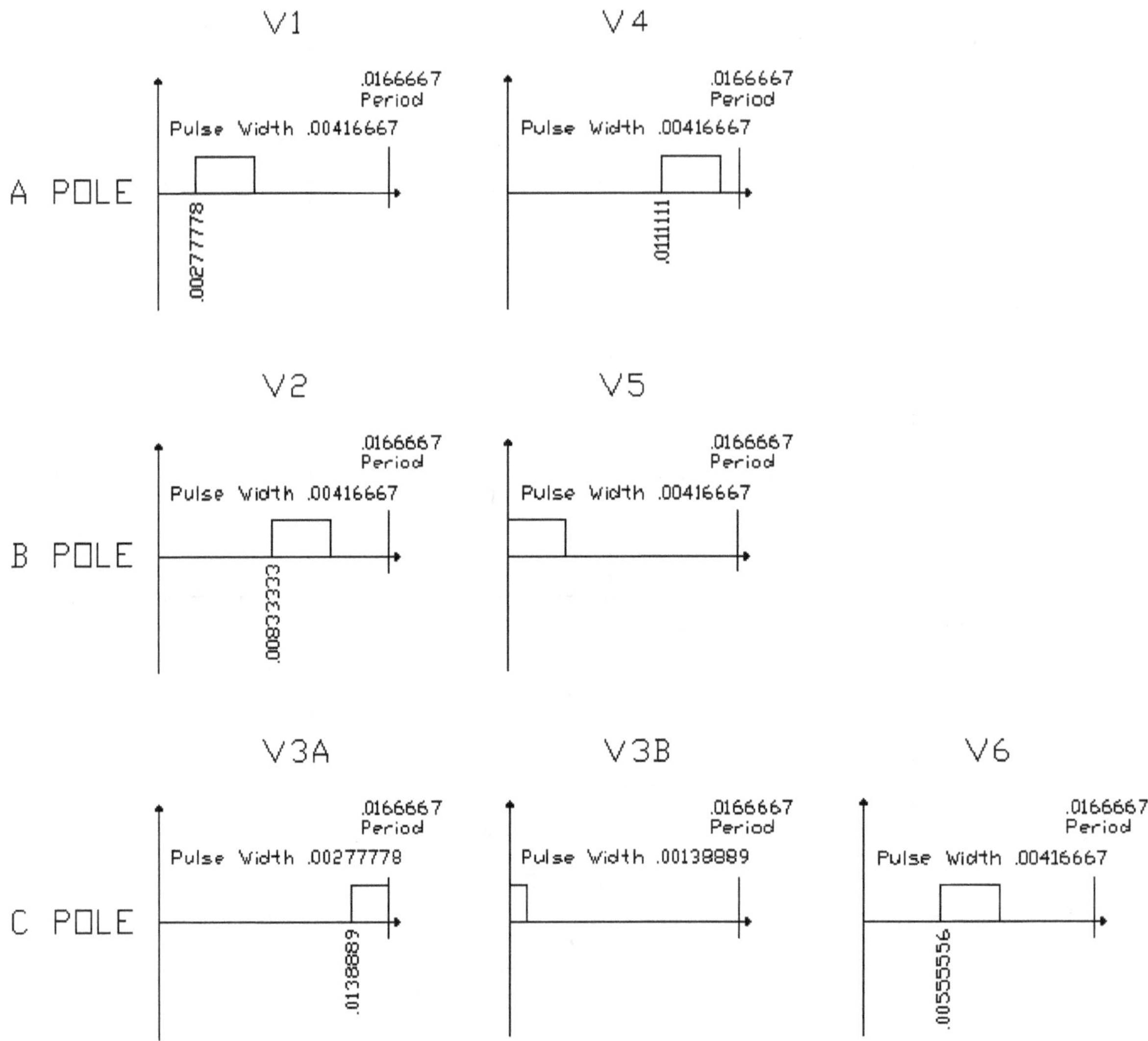

Figure 4-1-2-2 Switch pulse on diagrams. All use a period of .0166667 seconds, equivalent to a frequency of 60 Hz. The pulses turn on each switch for 90° of each 360° cycle.

6) Values for the supply voltage and components are:

Description	Symbol	Value and units
Fundamental AC sinusoidal voltage	VAS	0 V offset, 813.2 V_{ACpeak}, 60 Hz frequency, 0 second delay time, 0 sec^{-1} damping factor, and 0° phase angle. This is shown in Figure 4-1-1-3.
Fundamental AC sinusoidal voltage	VBS	0 V offset, 813.2 V_{ACpeak}, 60 Hz frequency, 0 second delay time, 0 sec^{-1} damping factor, and -120° phase angle. This is shown in Figure 4-1-1-3.
Fundamental AC sinusoidal voltage	VCS	0 V offset, 813.2 V_{ACpeak}, 60 Hz frequency, 0 second delay time, 0 sec^{-1} damping factor, and -240° phase angle. This is shown in Figure 4-1-1-3.
Positive notch voltage	VAU	0 V initial voltage, 1125 V pulsed voltage, .01202 second time delay, 0 second rise time, 0 second fall time, .0004 second pulse width, .0166667 second period. This is shown in Figure 4-1-1-3.
Positive notch voltage	VBU	0 V initial voltage, 1125 V pulsed voltage, .0009089 second time delay, 0 second rise time, 0 second fall time, .0004 second pulse width, .0166667 second period. This is shown in Figure 4-1-1-3.
Positive notch voltage	VCU	0 V initial voltage, 1125 V pulsed voltage, .00646 second time delay, 0 second rise time, 0 second fall time, .0004 second pulse width, .0166667 second period. This is shown in Figure 4-1-1-3.

Continued

Negative notch voltage	VAD	0 V initial voltage, -1125 V pulsed voltage, .00369 second time delay, 0 second rise time, 0 second fall time, .0004 second pulse width, .0166667 second period. This is shown in Figure 4-1-1-3.
Negative notch voltage	VBD	0 V initial voltage, -1125 V pulsed voltage, .00925 second time delay, 0 second rise time, 0 second fall time, .0004 second pulse width, .0166667 second period. This is shown in Figure 4-1-1-3.
Negative notch voltage	VCD	0 V initial voltage, -1125 V pulsed voltage, .01480 second time delay, 0 second rise time, 0 second fall time, .0004 second pulse width, .0166667 second period. This is shown in Figure 4-1-1-3.
Line resistance	RLA, RLB, RLC	.25 Ω
Line inductance	LLA, LLB, LLC	.0004 H
Diodes	DG1 to DG6	PLAIN diodes using the PSpice default parameters for simple ideal diodes
Capacitances	CS1 to CS6	.000006 F
Effective snubber circuit inductances	LS1 to LS6	.8E-6 H
Snubber resistor resistances	RS1 to RS6	.1 Ω
Switch circuit resistor resistances	R1 to R6	1000 Ω
Switches	S1 to S6	On resistance = 1E-3 Ω, Off resistance = 1E6 Ω, Turn-on voltage = .6 V, Turn-off voltage = .4 V
Switch turn-on voltage pulses	V1, V2, V3A, V3B, V4, V5, V6	.0166667 second period 1 V pulses of 90° widths to turn-on the switches. These are shown in Figure 4-1-2-2.
Equivalent motor inductance	LM	.0457 H
Equivalent motor resistance	RM	12.1 Ω

7) The netlist program is:

```
Figure 4-1-2-1.CIR CONVERTER AT HALF OUTPUT WITH NOTCHED INPUT VOLTAGE
* Plotting converter and load voltages versus time and finding the output
* form factor with SCRs that are on half the time
VAS 1 0 SIN(0 813.2 60 0 0 0)
VBS 2 0 SIN(0 813.2 60 0 0 -120)
VCS 3 0 SIN(0 813.2 60 0 0 -240)
*
VAU 4 1 PULSE(0 1125V .01202 0 0 .0004 .0166667)
VBU 5 2 PULSE(0 1125V .0009089 0 0 .0004 .0166667)
VCU 6 3 PULSE(0 1125V .00646 0 0 .0004 .0166667)
*
VAD 4 7 PULSE(0 1125V .00369 0 0 .0004 .0166667)
VBD 5 8 PULSE(0 1125V .00925 0 0 .0004 .0166667)
VCD 6 9 PULSE(0 1125V .0148 0 0 .0004 .0166667)
*
RLA 7 10 .25
RLB 8 11 .25
RLC 9 12 .25
*
LLA 10 13 .0004
LLB 11 14 .0004
LLC 12 15 .0004
*
DG1 13 20 PLAIN
DG2 14 22 PLAIN
DG3 15 24 PLAIN
DG4 17 21 PLAIN
DG5 17 23 PLAIN
DG6 17 25 PLAIN
.MODEL PLAIN D
*
CS1 27 26 .000006
CS2 29 28 .000006
CS3 31 30 .000006
CS4 33 32 .000006
CS5 35 34 .000006
CS6 37 36 .000006
*
```

```
LS1 26 16 .8E-6
LS2 28 16 .8E-6
LS3 30 16 .8E-6
LS4 32 13 .8E-6
LS5 34 14 .8E-6
LS6 36 15 .8E-6
*
RS1 13 27 .1
RS2 14 29 .1
RS3 15 31 .1
RS4 17 33 .1
RS5 17 35 .1
RS6 17 37 .1
*
V1 51 0 PULSE(0 1 .00277778 0 0 .00416667 .0166667)
V2 52 0 PULSE(0 1 .00833333 0 0 .00416667 .0166667)
V3A 53 57 PULSE(0 1 .0138889 0 0 .00277778 .0166667)
V3B 57 0 PULSE(0 1 0 0 0 .00138889 .0166667)
V4 54 0 PULSE(0 1 .0111111 0 0 .00416667 .0166667)
V5 55 0 PULSE(0 1 0 0 0 .00416667 .0166667)
V6 56 0 PULSE(0 1 .00555556 0 0 .00416667 .0166667)
*
R1 51 0 1000
R2 52 0 1000
R3 53 0 1000
R4 54 0 1000
R5 55 0 1000
R6 56 0 1000
*
S1 16 20 51 0 RELAY
S2 16 22 52 0 RELAY
S3 16 24 53 0 RELAY
S4 13 21 54 0 RELAY
S5 14 23 55 0 RELAY
S6 15 25 56 0 RELAY
.MODEL RELAY VSWITCH(RON=1E-3 ROFF=1E6 VON=.6 VOFF=.4)
*
LM 16 18 .0457
RM 16 17 12.1
*
.TRAN .1m 40m
.PROBE
.END
```

8) The voltage waveforms of Figures 4-1-1-5, 4-1-1-6, and 4-1-1-7 are repeated with the half-power output converter.

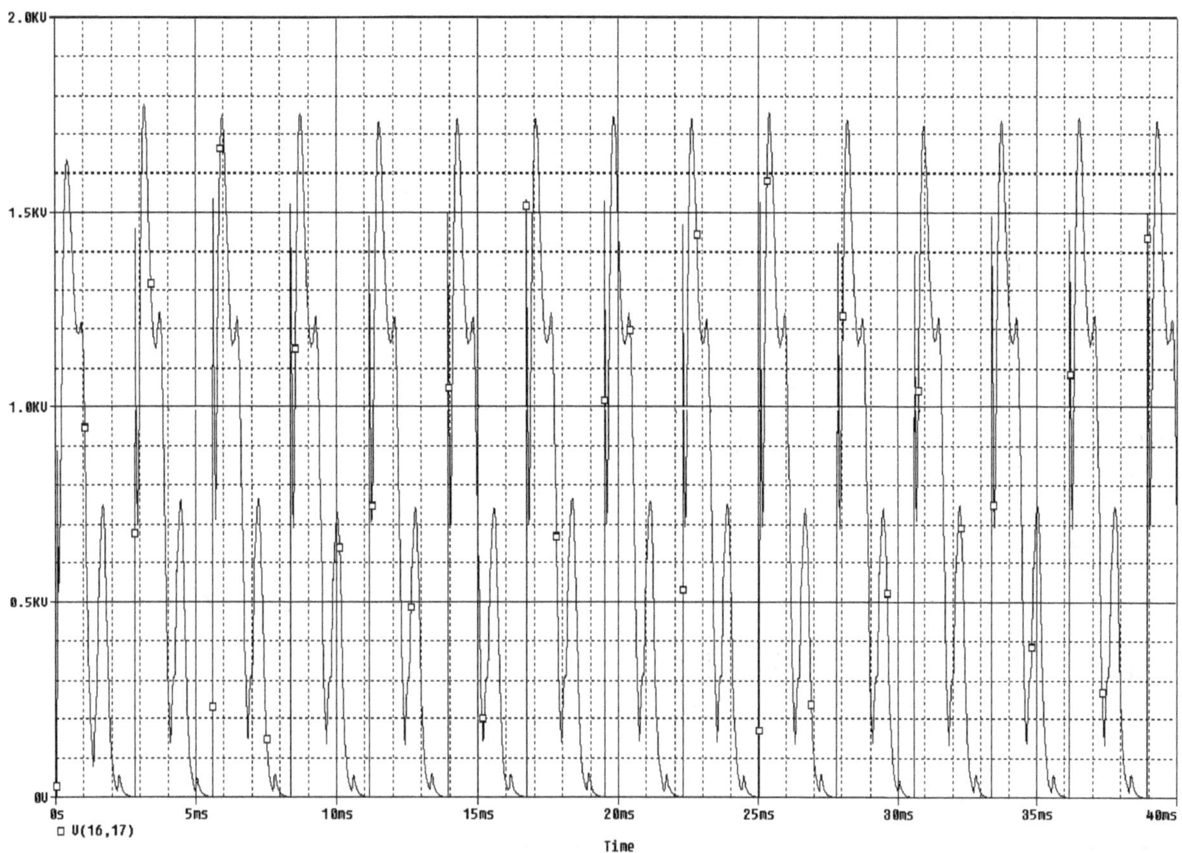

Figure 4-1-2-3 Voltage across the series motor.

9) As would be expected, the output voltage is much choppier than at full-load. The Probe AVG (average) and RMS graphs quantify the difference.

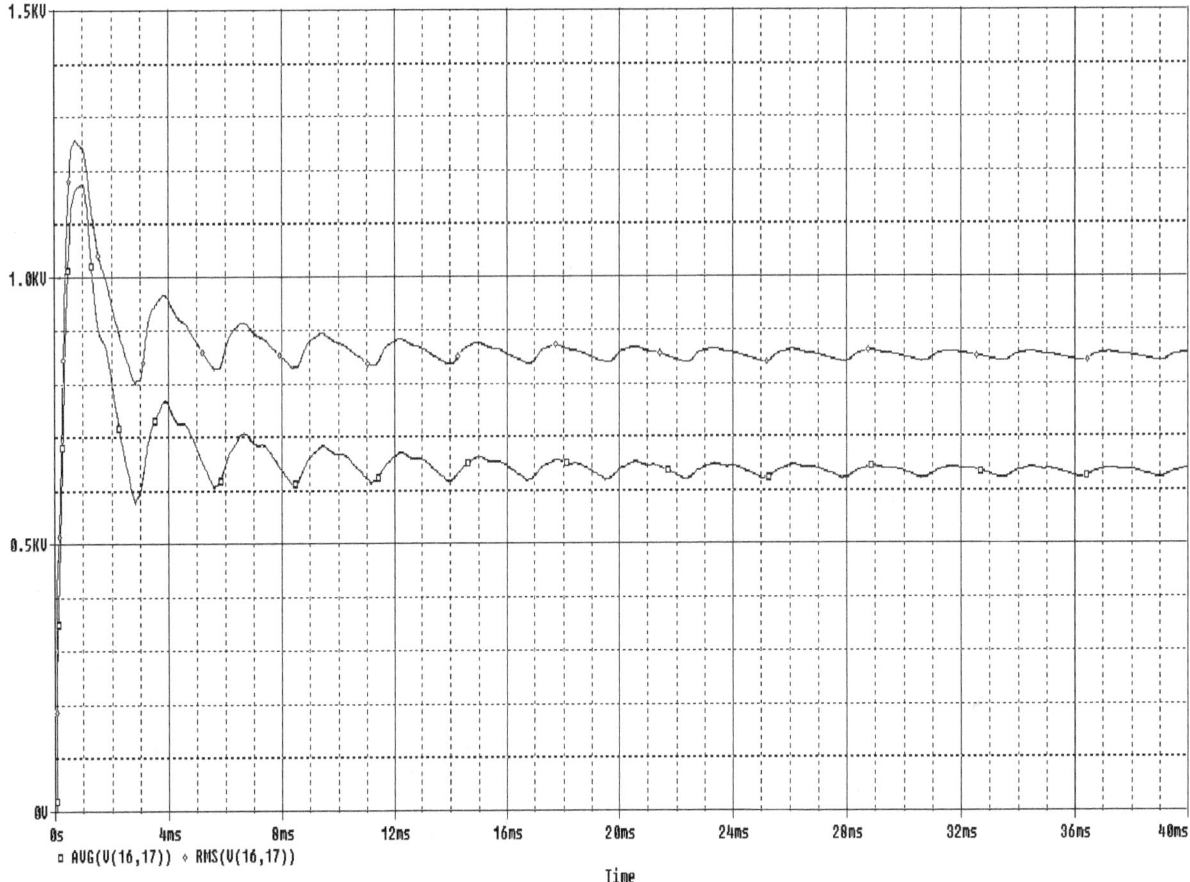

Figure 4-1-2-4 Average and rms values of the voltage across the series motor.

10) A Form Factor can be calculated for the voltage.

Form Factor = V_{ACrms}/V_{DC}

In this example at time 32 milliseconds, the Form Factor = .858/.642 = 1.34

This Form Factor can be compared to that of the series motor to determine if the motor is able to withstand AC riders of this magnitude. A normal DC series motors is not rated for a Form Factors of 1.34. However, “SCR rated” DC motors are rated to receive voltages with Form Factors of up to 1.3 to 1.4.

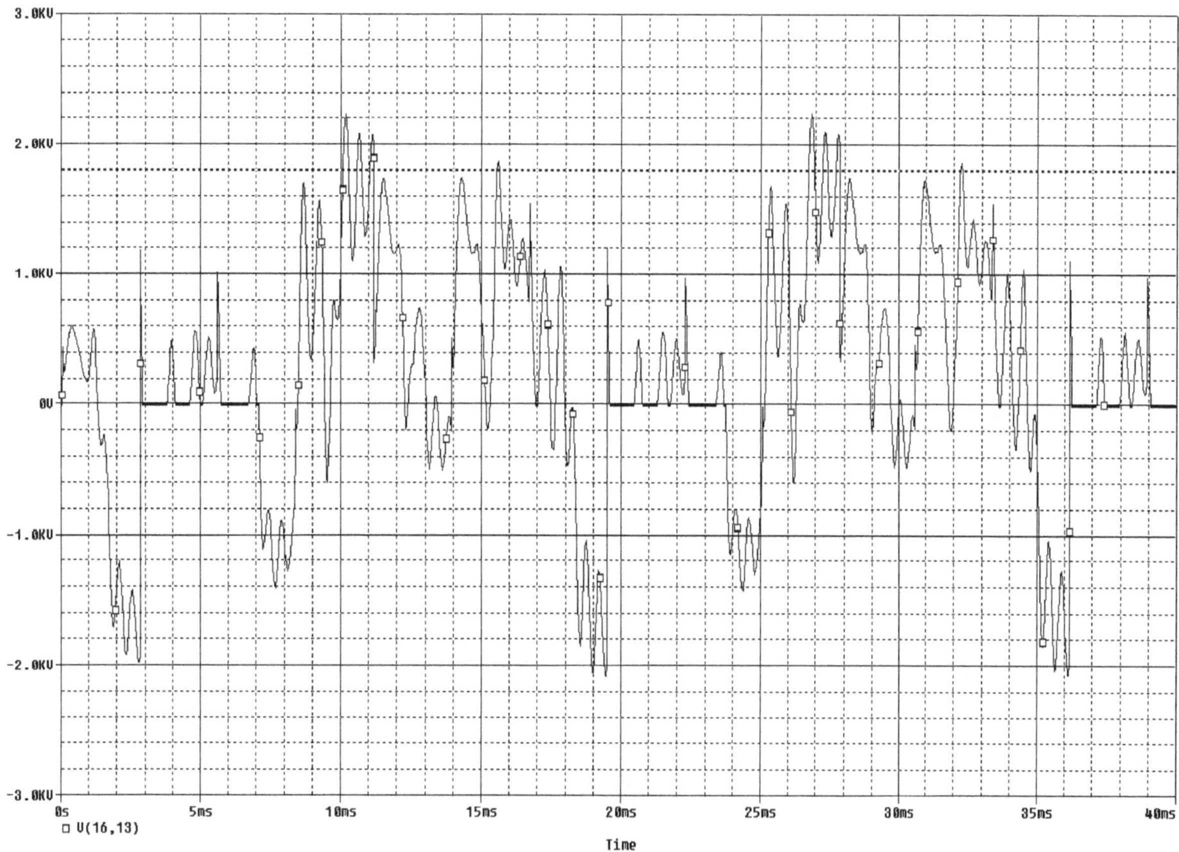

Figure 4-1-2-5 Voltage across SCR AP (diode DG1 in series with switch S1) during half-power output operation.

11) Thus far, the PSpice analysis has assumed that the SCRs stay on continuously after they have switched on. The switches used in the half-power analysis were directed to stay on continuously for one half of the full conduction cycle. However, if in the actual circuit the SCRs are only receiving a short gate pulse at the beginning of each conduction cycle, they would turn off if they received a reverse voltage. Looking at Figure 4-1-2-5, in the times where the SCR is in conduction, such as 20 milliseconds to 24 milliseconds, shows that reverse voltage spikes are occurring across the SCRs. If an SCR received such spikes and did not have its gating signal on continuously, it would turn off and stay off. The turning off and staying off of the SCRs would result in an even choppier output voltage than that shown in Figure 4-1-2-3.

12) An attempt to reduce the effect of notch depth can be made by placing increased inductance in the line to the converter. The netlist program could be run with increased inductance for LLA, LLB, and LLC to simulate series line inductors. The choppiness that occurs during reduced output could be decreased by improving the snubber design. Each modification could be tested and optimized with PSpice without building an actual circuit.

4.2 SIX-STEP INVERTER

PROBLEM: A subway car's propulsion inverter is powered by DC voltage from a third rail. The propulsion inverter then supplies three-phase induction motors to drive the railcar. The inverter produces a six-step (simple square-wave) three-phase. What are the magnitudes and frequencies of the AC voltage that the inverter sends back to the third rail? What is the effect of doubling the inverter's input capacitor capacitance?

DISCUSSION:

There are many propulsion inverter circuits used on subway cars. The oldest use SCRs, newer ones use GTOs, and currently IGBTs are in common use. This circuit uses GTOs and simple standard resistor/capacitor/diode snubbers. Most inverters would use more complicated snubber circuits.

Only 60 Hz will be produced by the inverter in this example. Actual traction motor inverters are designed to produce a range of frequencies. A more extensive analysis would look at noise production at more frequencies.

The three-phase induction motors are represented by equivalent circuits of resistors in series with inductors. This representation is good for steady-state operation. If the motors' speed and output power were changed to other steady-state values, the values of the resistors and inductors would be changed to agree with them. If the motors were changing speed, as they might be during startup or with a changing load, a dynamic circuit model should be used. Without the addition of the SLPS Interface (**S**imu**L**ink **PS**pice Interface), such dynamic circuit models are beyond the capabilities of PSpice

PROCEDURE:

1) Draw the basic power circuit.

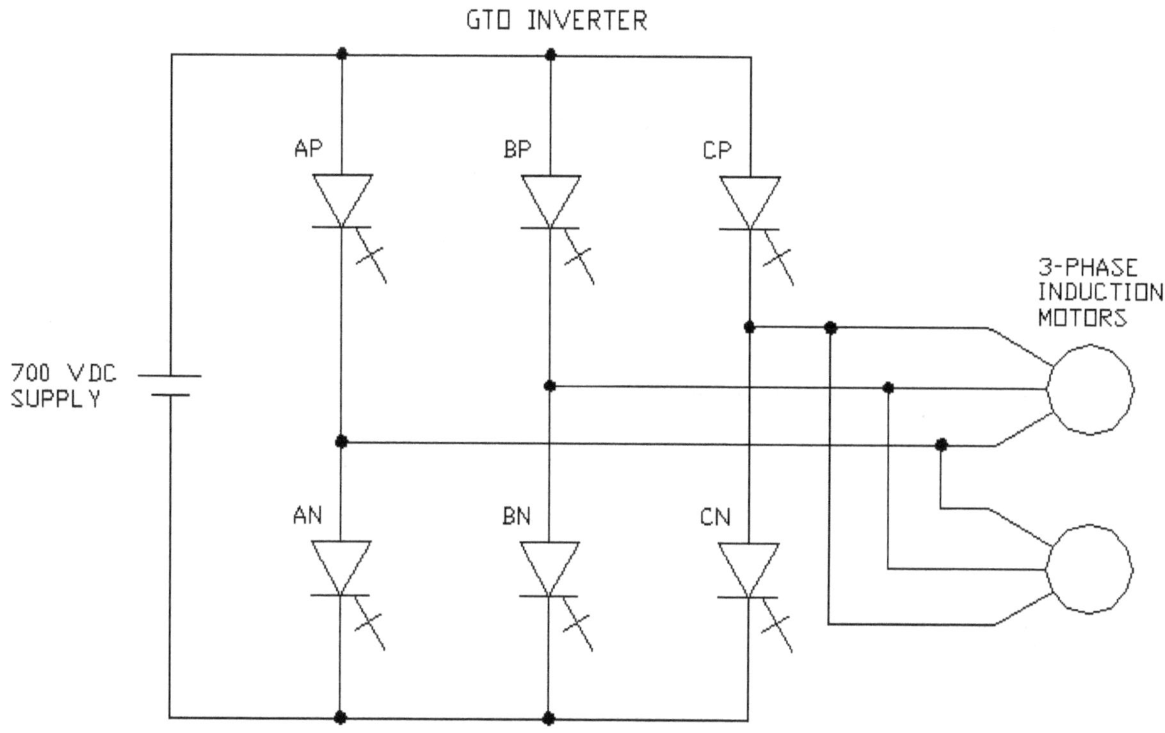

Figure 4-2-1 Basic power circuit of a GTO six-step inverter receiving DC power from the third rail and supplying AC power to two railcar traction motors.

2) Redraw the circuit in more detail in a format that can be analyzed by PSpice. The third rail supply is connected through an effective rail inductance and resistance. An input filtering circuit, series reactor, and charging capacitor are added to the input of the inverter. The GTOs are replaced with controlled switches in series with diodes. Each controlled switch is turned on for 120^0 of every 360^0. Freewheeling diodes and resistor/capacitor/diode snubbers are connected across each GTO equivalent. Snubber circuits include inductances that approximate the inductance of each snubber circuit. The two traction motors are assumed to be in steady-state operation. In steady-state operation they are simplified to one equivalent series resistance and inductance per phase.

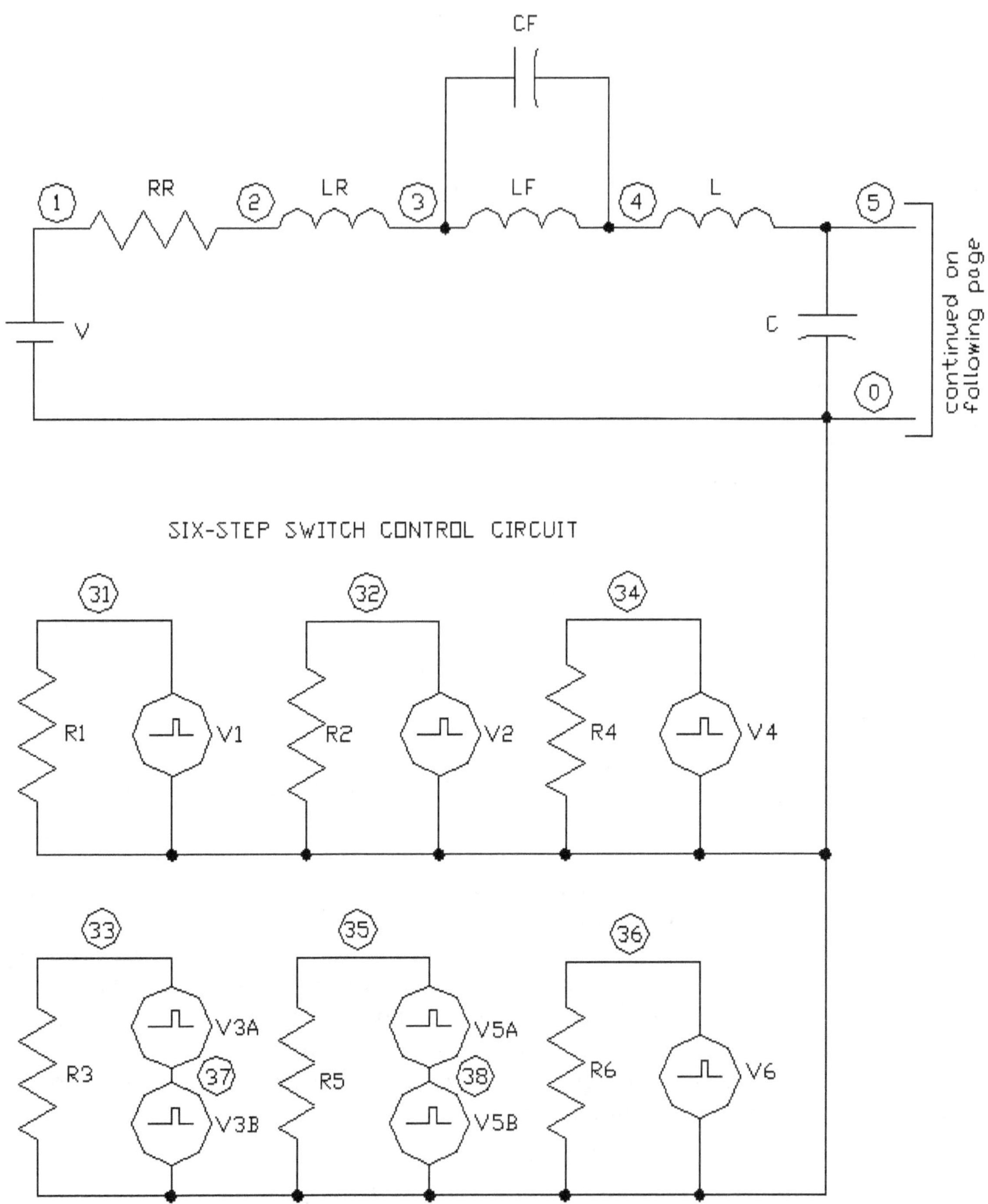
RAIL SUPPLY AND INVERTER INPUT CIRCUIT
CF
RR
LR
LF
L
1
2
3
4
5
V
C
0
continued on following page
SIX-STEP SWITCH CONTROL CIRCUIT
31
32
34
R1
V1
R2
V2
R4
V4
33
35
36
R3
V3A
37
V3B
R5
V5A
38
V5B
R6
V6

Continued

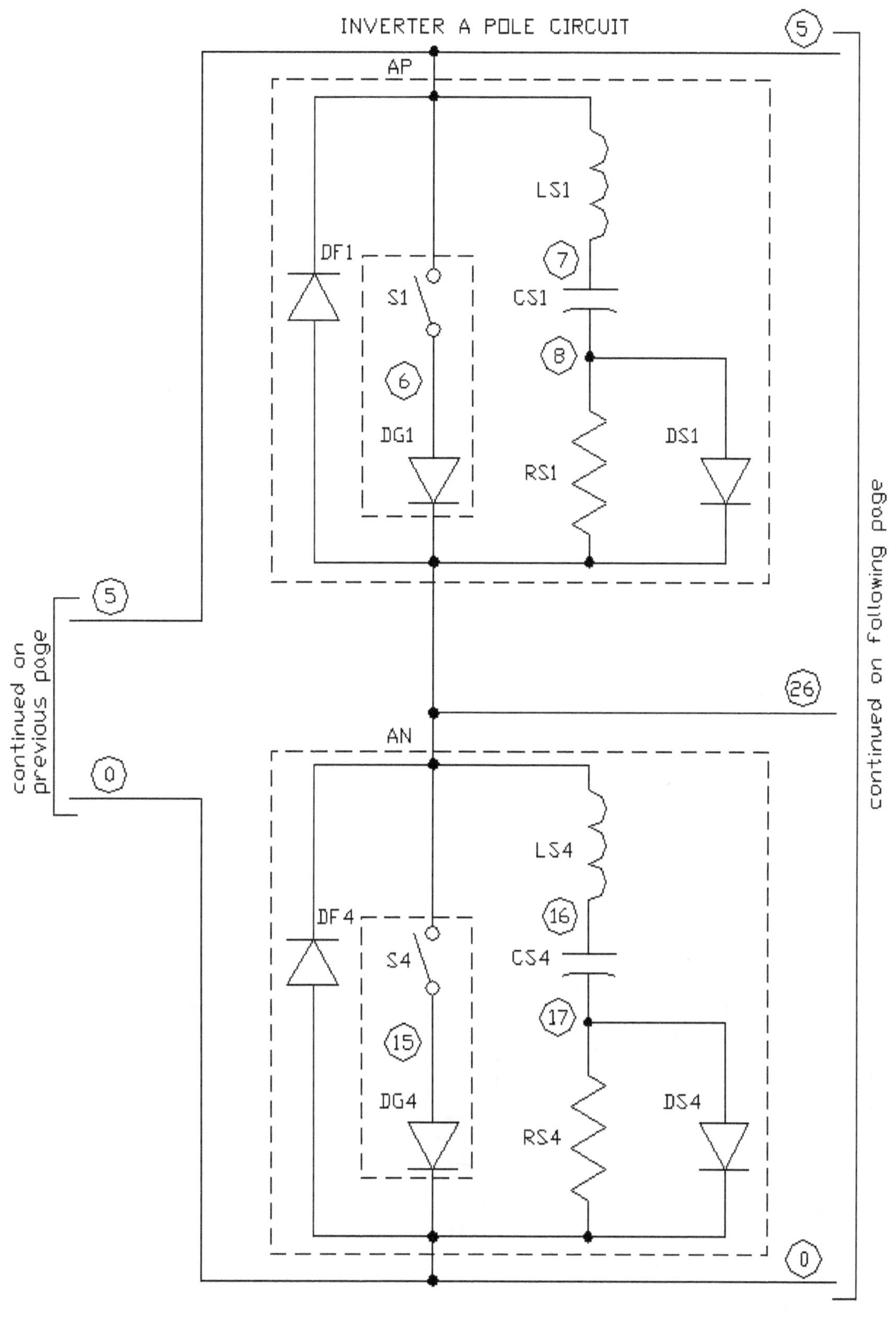
INVERTER A POLE CIRCUIT
5
AP
LS1
DF1
7
S1
CS1
8
6
DG1
DS1
RS1
5
continued on previous page
26
continued on following page
AN
0
LS4
16
DF4
S4
CS4
17
15
DG4
DS4
RS4
0

Continued

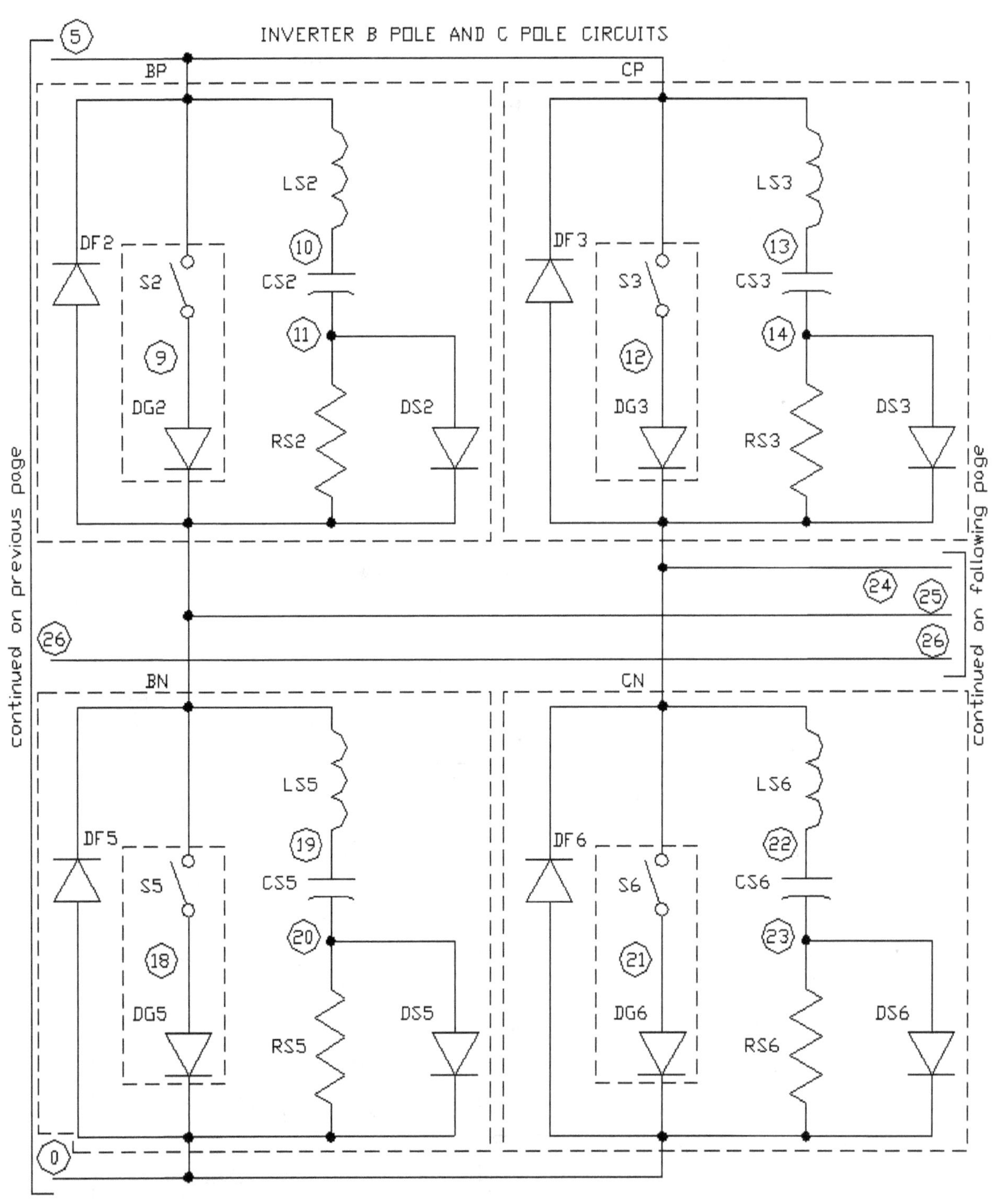

INVERTER B POLE AND C POLE CIRCUITS
5
BP
CP
LS2
LS3
DF2
DF3
10
13
S2
S3
CS2
CS3
11
14
9
12
DG2
DS2
DG3
DS3
RS2
RS3
24
25
26
26
BN
CN
LS5
LS6
DF5
DF6
19
22
S5
S6
CS5
CS6
20
23
18
21
DG5
DS5
DG6
DS6
RS5
RS6
0
continued on previous page
continued on following page

Continued

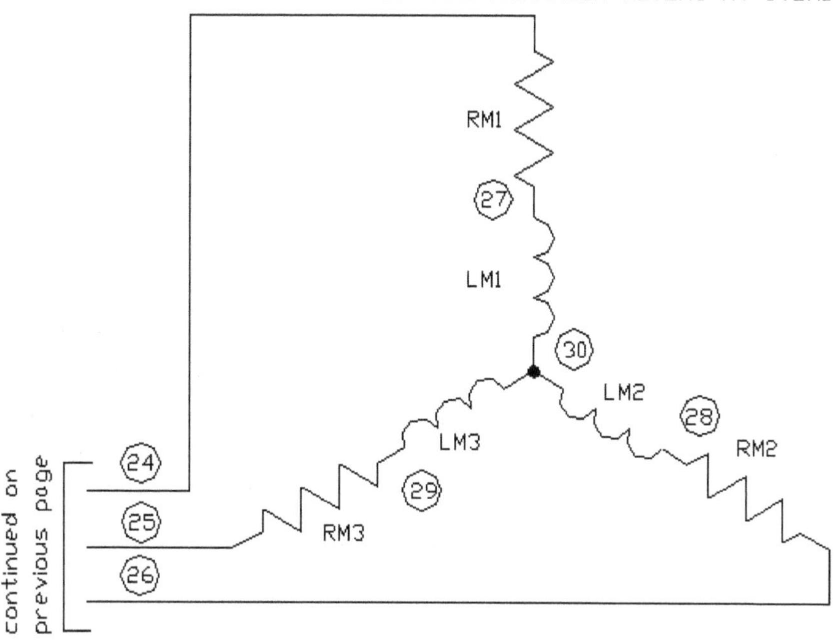

Figure 4-2-2 Schematic of a GTO inverter receiving power from a third rail and supplying power to two railcar traction motors.

3) Values for the supply voltage and components are:

Description	Symbol	Value and units
Supply voltage	V	700 V_{DC}
Rail resistance	RR	.00863 Ω
Rail inductance	LR	1E-9 H
Filter capacitor capacitance	CF	.001E-6 F
Filter inductor inductance	LF	.0004 H
Line reactor inductance	L	.0008 H
Charging capacitor capacitances	C	.187 or .376 F
Diodes	DF1 to DF6 DG1 to DG6 DS1 to DS6	Effective on series resistance = .03 Ω, Breakdown Voltage = 10000 V_{DC}, Current at Breakdown Voltage = 50 milliamps
Capacitances	CS1 to CS6	.000006 F

Continued

Effective snubber circuit inductances	LS1 to LS6	.8E-6 H
Snubber resistor resistances	RS1 to RS6	.1 Ω
Motor equivalent resistances	RM1 to RM3	.222 Ω
Motor equivalent inductances	LM1 to LM3	.000589 H
Switch circuit resistor resistances	R1 to R6	1000 Ω
Switches	S1 to S6	On resistance = 1E-3 Ω, Off resistance = 1E6 Ω, Turn-on voltage = .6 V, Turn-off voltage = .4 V
Switch turn-on voltage pulses	V1, V2, V3A, V3B, V4, V5A, V5B, V6	.0166667 second period 1 V pulses of varying widths to turn on the switches. These are shown in Figure 4-2-3.

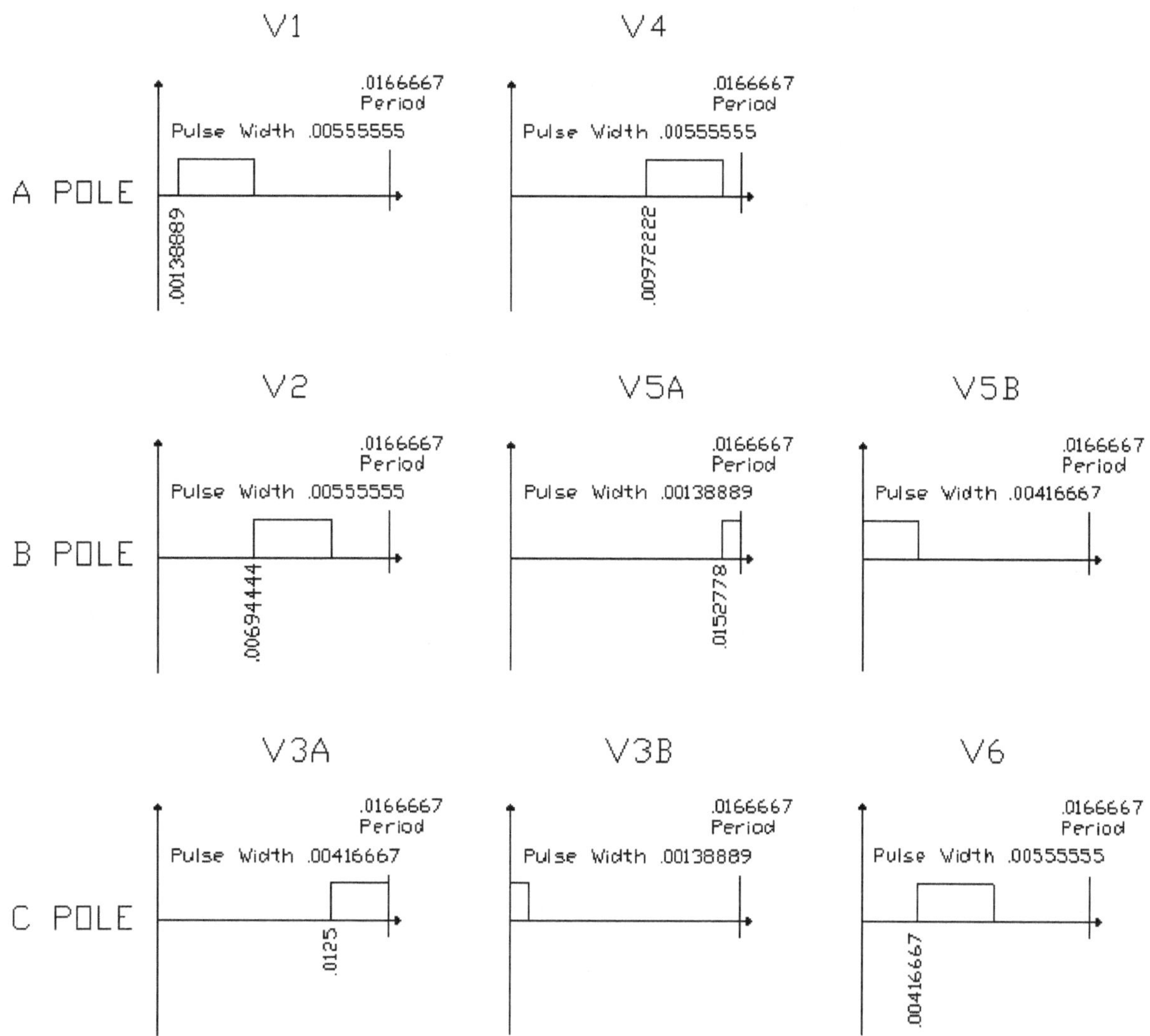

Figure 4-2-3 Switch pulse on diagrams. All use a period of .0166667, equivalent to a frequency of 60 Hz. The pulses turn on each switch for 120° of each 360° cycle.

4) Write the netlist program:

```
Figure 4-2-1.CIR SQUARE-WAVE SIX-STEP INVERTER
* Plotting motor input current and voltage and voltage at the inverter
* input versus time. Two charging capacitances are used.
*
V 1 0 700
*
RR 1 2 .00863
*
LR 2 3 1E-9
*
CF 3 4 .001E-6
*
LF 3 4 .0004
*
L 4 5 .0008
*
C 0 5 .187
*The value of C is set at l87 and .374 in different program runs
*
DF1 26 5 TYPE1
DF2 25 5 TYPE1
DF3 24 5 TYPE1
DF4 0 26 TYPE1
DF5 0 25 TYPE1
DF6 0 24 TYPE1
*
DG1 6 26 TYPE1
DG2 9 25 TYPE1
DG3 12 24 TYPE1
DG4 15 0 TYPE1
DG5 18 0 TYPE1
DG6 21 0 TYPE1
*
DS1 8 26 TYPE1
DS2 11 25 TYPE1
DS3 14 24 TYPE1
DS4 17 0 TYPE1
DS5 20 0 TYPE1
DS6 23 0 TYPE1
*
.MODEL TYPE1 D(RS=.03 BV=10000 IBV=50m)
```

```
*
CS1 7 8 .000006
CS2 10 11 .000006
CS3 13 14 .000006
CS4 16 17 .000006
CS5 19 20 .000006
CS6 22 23 .000006
*
LS1 5 7 .8E-6
LS2 5 10 .8E-6
LS3 5 13 .8E-6
LS4 26 16 .8E-6
LS5 25 19 .8E-6
LS6 24 22 .8E-6
*
RS1 8 26 .1
RS2 11 25 .1
RS3 14 24 .1
RS4 17 0 .1
RS5 20 0 .1
RS6 23 0 .1
*
RM1 24 27 .222
RM2 26 28 .222
RM3 25 29 .222
*
LM1 30 27 .000589
LM2 30 28 .000589
LM3 30 29 .000589
*
R1 31 0 1000
R2 32 0 1000
R3 33 0 1000
R4 34 0 1000
R5 35 0 1000
R6 36 0 1000
```

```
*
S1 5 6 31 0 RELAY
S2 5 9 32 0 RELAY
S3 5 12 33 0 RELAY
S4 26 15 34 0 RELAY
S5 25 18 35 0 RELAY
S6 24 21 36 0 RELAY
*
.MODEL RELAY VSWITCH(RON=1E-3 ROFF=1E6 VON=.6 VOFF=.4)
*
V1 31 0 PULSE(0 1 .00138889 0 0 .00555555 .0166667)
V2 32 0 PULSE(0 1 .00694444 0 0 .00555555 .0166667)
V3A 33 37 PULSE(0 1 .0125 0 0 .00416667 .0166667)
V3B 37 0 PULSE(0 1 0 0 0 .00138889 .0166667)
V4 34 0 PULSE(0 1 .00972222 0 0 .00555555 .0166667)
V5A 35 38 PULSE(0 1 .0152778 0 0 .00138889 .0166667)
V5B 38 0 PULSE(0 1 0 0 0 .00416667 .0166667)
V6 36 0 PULSE(0 1 .00416667 0 0 .00555555 .0166667)
*
.TRAN .000006 2
*
.PROBE
.END
```

5) Once the simulation is run, Probe can be used to display traces of voltages and currents throughout the circuit.

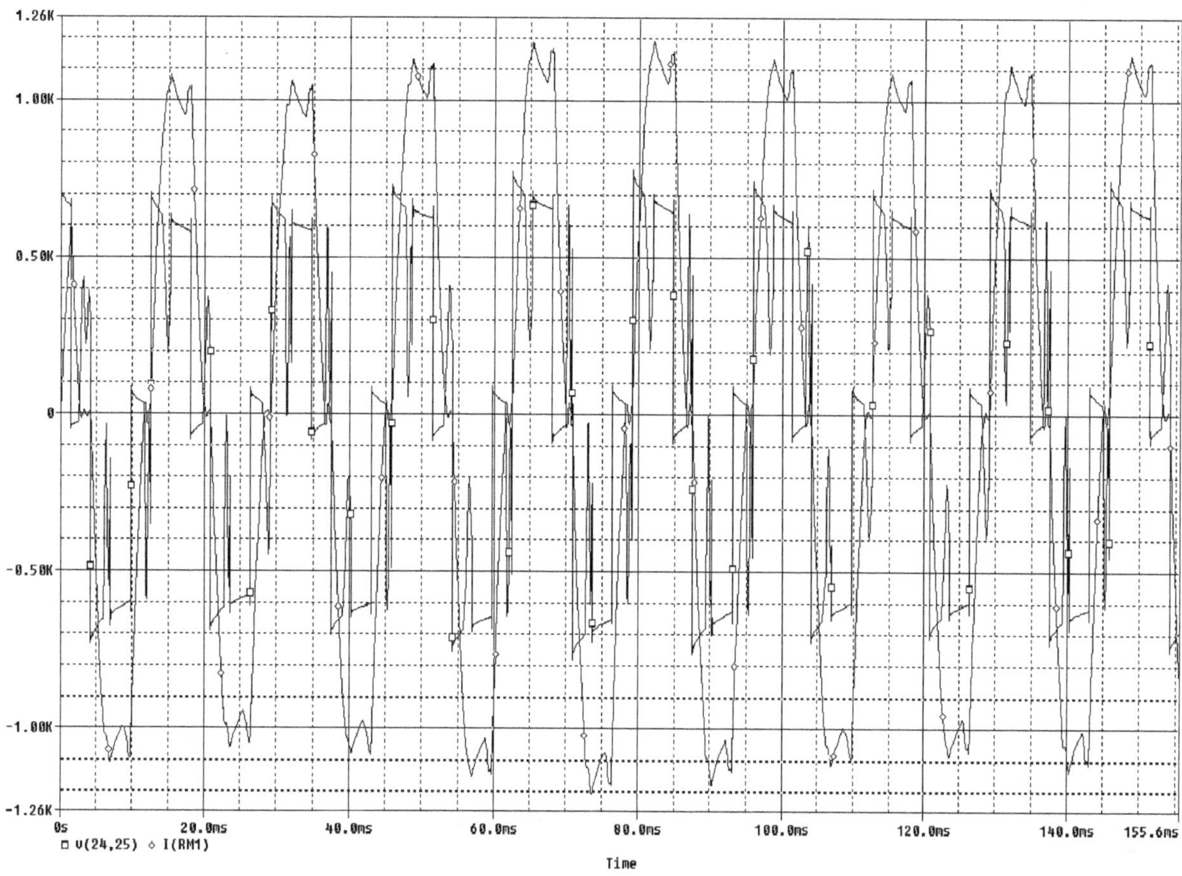

Figure 4-2-4 Line-to-line voltage applied to the motors and current through one phase of the motors. Probe "Zoom Area" has been used to display this region of the output. C equals .187 F.

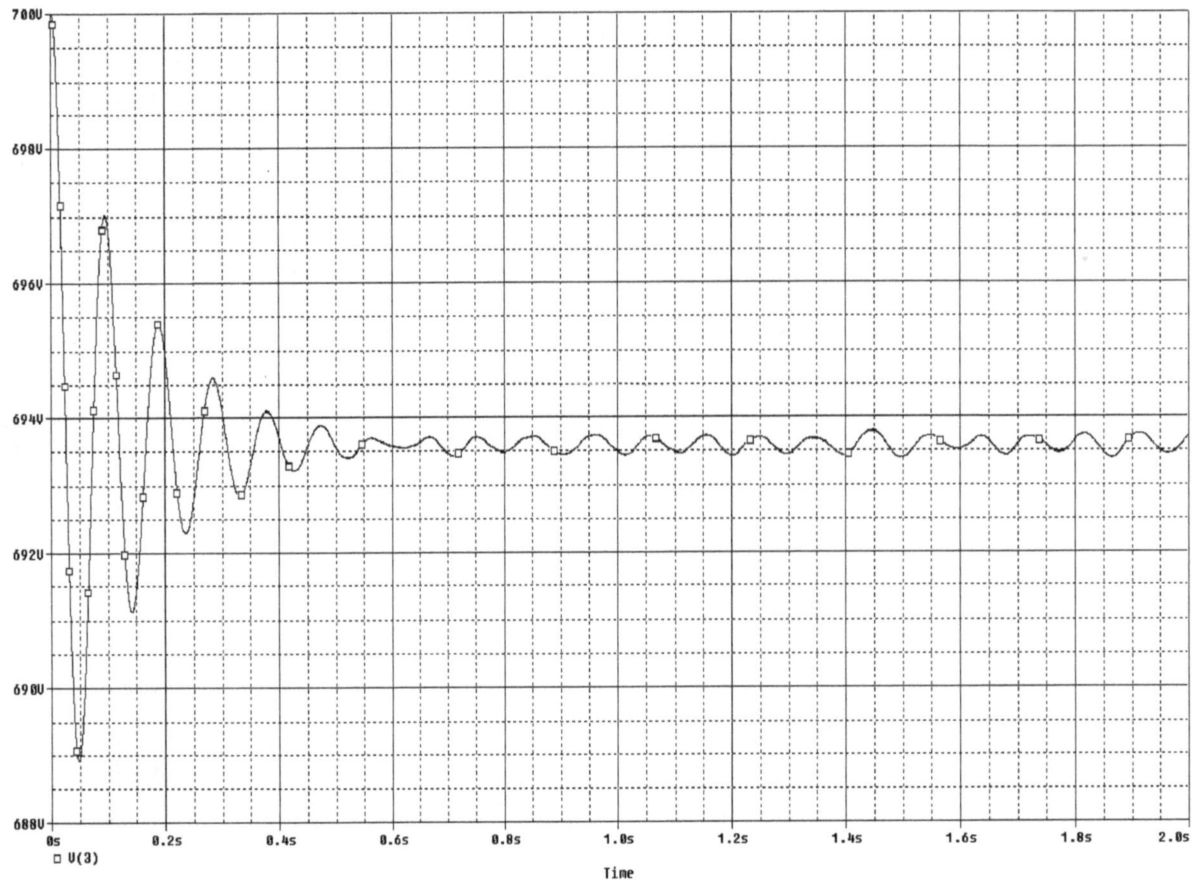

Figure 4-2-5 Voltage at the input to the inverter. C equals .187 F.

6) Figure 4-2-5 shows the AC noise that the inverter injects onto the DC bus. After .8 seconds, the noise voltage has a primary frequency of about 10 Hz and a peak value of about .15 volts.

7) The value of the input capacitor, C, is doubled to become .374 F and the netlist program rerun. The voltage at the input to the inverter is again displayed with Probe. The netlist program was again run for 2 seconds and the same time window displayed in Figure 4-2-6.

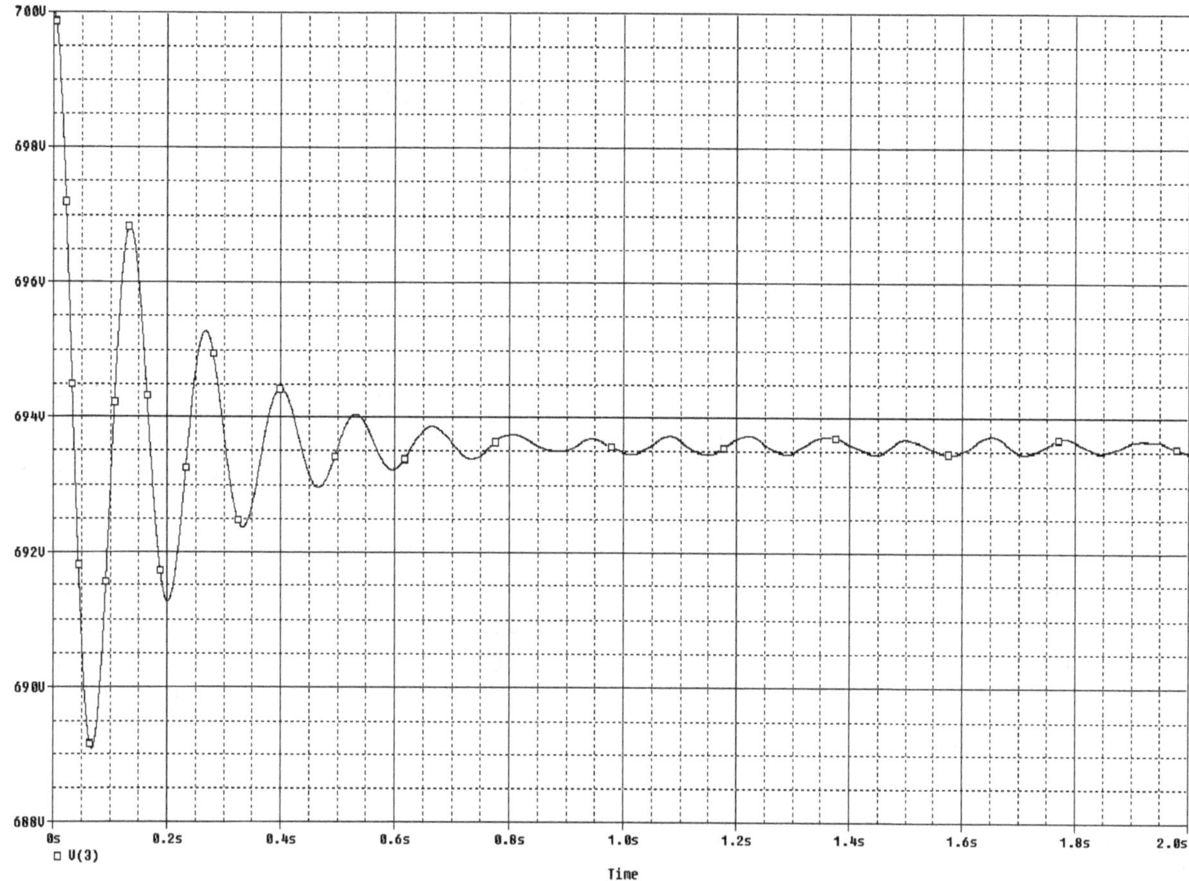

Figure 4-2-6 Voltage at the input to the inverter with the input capacitor's capacitance, C, doubled to .374 F.

8) Figure 4-2-6 shows that the AC noise the inverter injects onto the DC bus. After .8 seconds, the noise voltage has a primary frequency of about 7 Hz and a peak value of about .12 volts.

9) The circuit and its netlist program can be modified in many ways. Then PSpice and Probe can be used to quickly show the results of the modifications.

4.3 PWM INVERTERS

PROBLEM: The same problem considered in Section 4.2 will be redone with PWM IGBT inverters. How do the waveforms of the voltage and current received by the induction motors look? What are the magnitudes and frequencies of the AC voltage that the inverter sends back to the third rail? How do these results compare with those of Section 4.2. Consider only the C equals .187 F case.

DISCUSSION:

PWM (**P**ulse **W**idth **M**odulation) converts DC to AC by rapidly switching on and off varying time durations (widths) of positive and negative polarity DC to the AC output load. The quality of the produced AC is generally better than that produced by the six-step method of Section 4.2.

There are three ways of producing PWM switching control signals. The oldest method uses the comparison of a triangle or saw-tooth wave to a reference sine wave. The look-up method uses a stored pulse pattern. The look-up method makes is possible to select pulse widths that avoid specific output harmonics. The space-vector method uses the comparison of measured stator flux to a reference. The triangle/sine wave reference method and look-up method will be used here.

PROCEDURE:

1) Draw the basic power circuit.

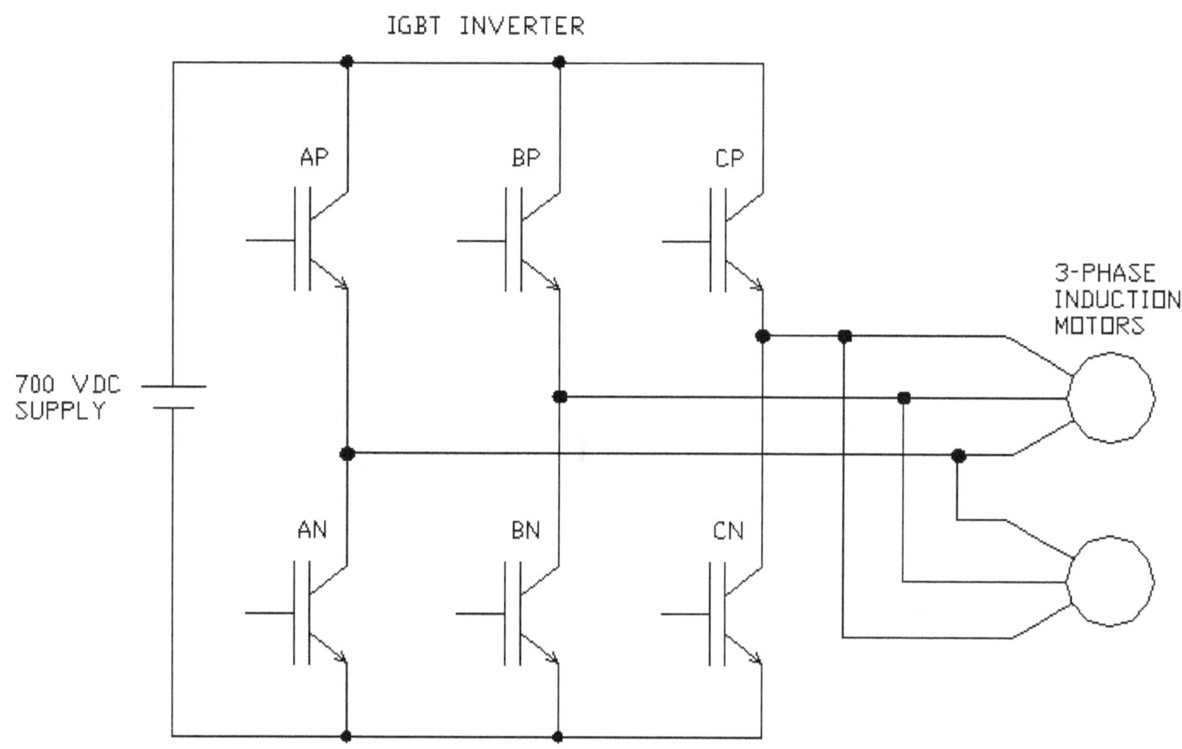

Figure 4-3-1 Basic power circuit of an IGBT PWM inverter receiving DC and producing AC.

2) Redraw the circuit in more detail in a format that can be analyzed by PSpice. The supply is connected through an effective inductance and resistance. An input filtering circuit, series reactor, and charging capacitor are added to the input of the inverter. The IGBTs are replaced with controlled switches in series with diodes. Each controlled switch is turned on and off many times each cycle. Freewheeling diodes are connected across each IGBT equivalent. Resistor/capacitor/diode snubbers are placed across each IGBT equivalent. The snubber circuits include inductances that approximate the inductance of each snubber circuit. The two traction motors are assumed to be in steady-state operation. They are reduced to one equivalent series resistance and inductance circuit.

4.3.1 PWM INVERTER WITH TRIANGLE/SINE WAVE CONTROL

The PSpice schematic for a PWM inverter with triangle/sine wave control is the same as that for the six-step inverter except that a triangle/sine wave control is used rather than a six-step control. Replace the six-step switch control circuit of Figure 4-2-2 with the triangle/sine wave PWM switch control circuit of Figure 4-3-1-1 to create the schematic.

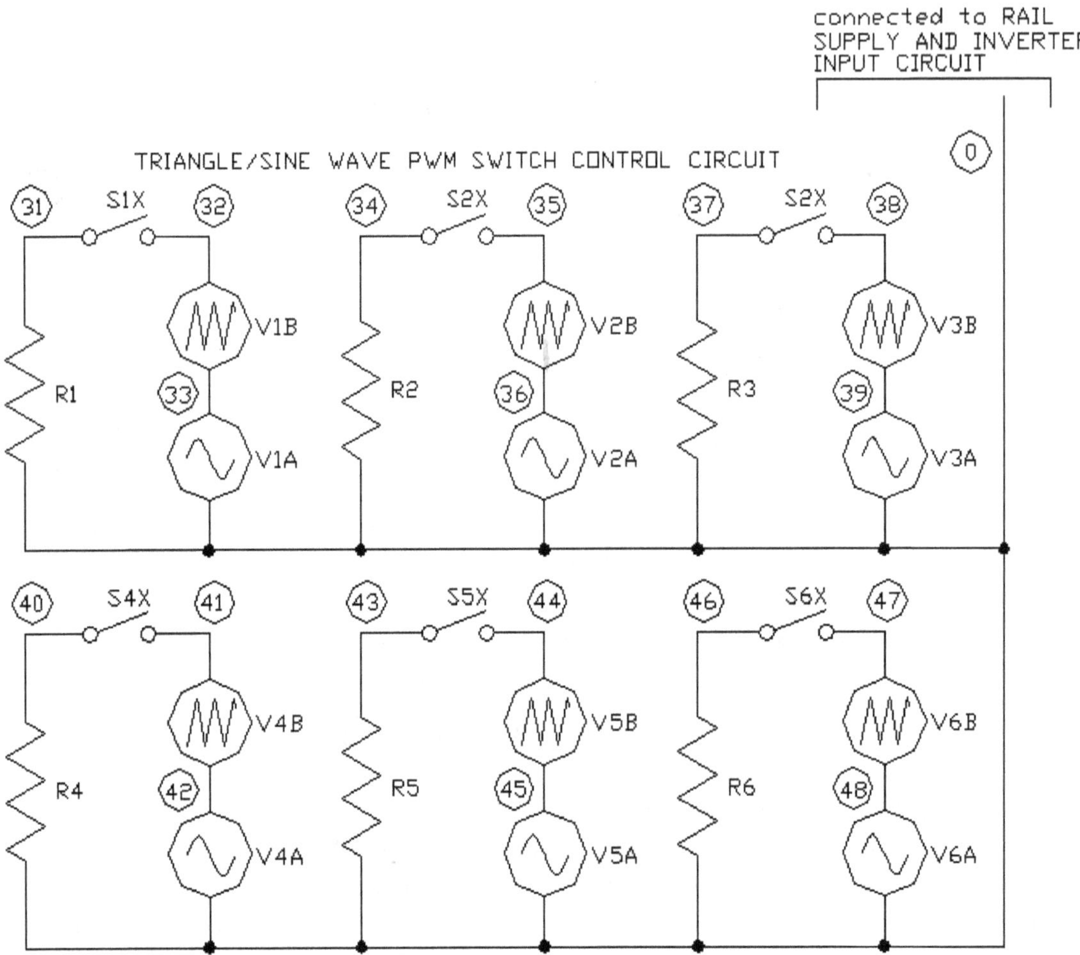

Figure 4-3-1-1 Control circuit of a triangle/sine wave controlled PWM IGBT inverter receiving power from a third rail and supplying power to two railcar traction motors. The rest of the schematic is the same as in Figure 4-2-2.

When the resistors of Figure 4-3-1-1 receive positive voltage their corresponding switches (the ones representing IGBTs) close.

In Figure 4-3-1-2 the resistor, R1, will receive positive voltage when the sine wave voltage, V1A, is greater than the triangular carrier wave voltage, V1B, and the voltage across V1A is greater than its midpoint voltage. The switch, S1X, is used to assure that the R1 resistor does not receive a positive voltage during the portion of the sine wave that is below its midpoint voltage.

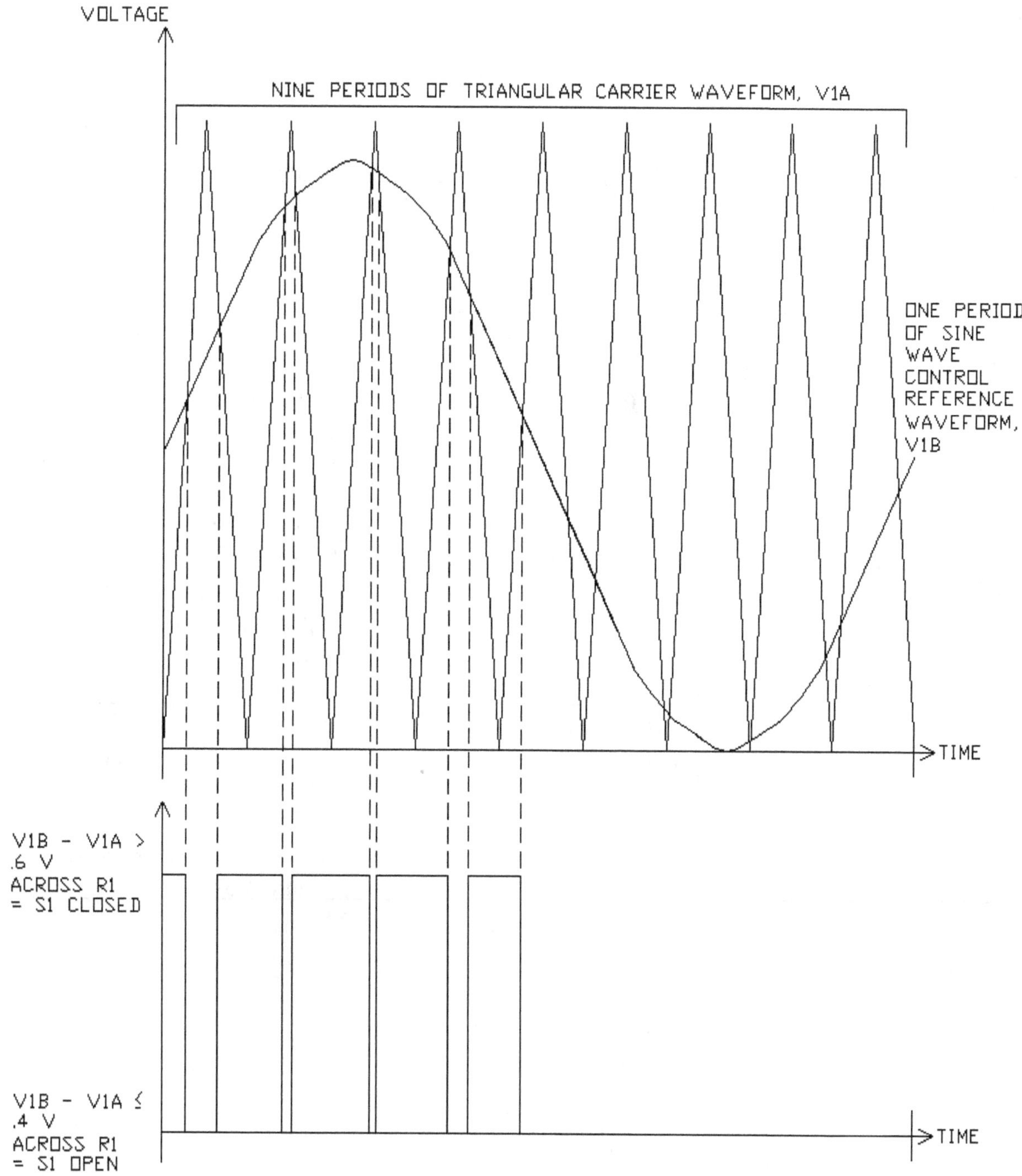

Figure 4-3-1-2 The reference sine wave voltage V1B, triangular carrier wave voltage V1A, and resulting voltage across the resistance R1 for the AP pole.

If switch S1 is already open, the voltage across R1 will close switch S1 whenever it is greater than .6 volts. If switch S1 is already closed, the S1 switch will open whenever the voltage across R1 is less than .4 volts.

PROCEDURE:

1) Values for the supply voltage and components are the same as those in Section 4.2 except:

DELETE THE FOLLOWING:

Description	**Symbol**	**Value and units**
Switch turn-on voltage pulses	V1, V2, V3A, V3B, V4, V5A, V5B, V6	.0166667 second period 1 V pulses of varying widths to turn on the switches. These are shown in Figure 4-2-3.

ADD THE FOLLOWING:

Description	**Symbol**	**Value and units**
Switches	S1X to S6X	On resistance = 1E-3 Ω, Off resistance = 1E6 Ω, Turn-on voltage = 100.4 V, Turn-off voltage = 100.1 V
Switch triangular carrier voltages	V1A, V2A, V3A, V4A, V5A, V6A	.00185185 second period 214 V_{ACpeak} triangle waveforms. All of these have the same waveform. These are shown in Figure 4-3-1-2.
Switch sine wave reference voltages	V1B, V2B, V3B, V4B, V5B, V6B	.0166667 second period, 100 V_{ACpeak}, 100 V offset (average DC) sine wave with phase shifts of 0, 120, 240, 180, 300, 60 degrees. V1B is shown in Figure 4-3-1-2.

2) Write the netlist program:

```
Figure 4-3-1a.CIR PWM CONTROLLED BY A TRIANGULAR WAVE AND A SINE WAVE
REFERENCE
* Plotting motor input current and voltage and voltage at the inverter
* input versus time.
*
V 1 0 700
*
RR 1 2 .00863
LR 2 3 1E-9
*
CF 3 4 .001E-6
LF 3 4 .0004
*
L 4 5 .0008
C 0 5 .187
*
DF1 26 5 TYPE1
DF2 25 5 TYPE1
DF3 24 5 TYPE1
DF4 0 26 TYPE1
DF5 0 25 TYPE1
DF6 0 24 TYPE1
*
DG1 6 26 TYPE1
DG2 9 25 TYPE1
DG3 12 24 TYPE1
DG4 15 0 TYPE1
DG5 18 0 TYPE1
DG6 21 0 TYPE1
*
DS1 8 26 TYPE1
DS2 11 25 TYPE1
DS3 14 24 TYPE1
DS4 17 0 TYPE1
DS5 20 0 TYPE1
DS6 23 0 TYPE1
*
.MODEL TYPE1 D(RS=.03 BV=10000 IBV=50m)
```

```
*
CS1 7 8 .000006
CS2 10 11 .000006
CS3 13 14 .000006
CS4 16 17 .000006
CS5 19 20 .000006
CS6 22 23 .000006
*
LS1 5 7 .8E-6
LS2 5 10 .8E-6
LS3 5 13 .8E-6
LS4 26 16 .8E-6
LS5 25 19 .8E-6
LS6 24 22 .8E-6
*
RS1 8 26 .1
RS2 11 25 .1
RS3 14 24 .1
RS4 17 0 .1
RS5 20 0 .1
RS6 23 0 .1
*
RM1 24 27 .222
RM2 26 28 .222
RM3 25 29 .222
*
LM1 30 27 .000589
LM2 30 28 .000589
LM3 30 29 .000589
*
R1 31 0 1000
R2 34 0 1000
R3 37 0 1000
R4 40 0 1000
R5 43 0 1000
R6 46 0 1000
```

```
*
S1 5 6 31 0 RELAY
S2 5 9 34 0 RELAY
S3 5 12 37 0 RELAY
S4 26 15 40 0 RELAY
S5 25 18 43 0 RELAY
S6 24 21 46 0 RELAY
*
.MODEL RELAY VSWITCH(RON=1E-3 ROFF=1E6 VON=.6 VOFF=.4)
*
V1A 33 0 SIN(100 100 60 0 0 0)
V2A 36 0 SIN(100 100 60 0 0 120)
V3A 39 0 SIN(100 100 60 0 0 240)
V4A 42 0 SIN(100 100 60 0 0 180)
V5A 45 0 SIN(100 100 60 0 0 300)
V6A 48 0 SIN(100 100 60 0 0 60)
*
V1B 33 32 PULSE(0 214 0 .000925925 .000925925 .00000001 .001851851)
V2B 36 35 PULSE(0 214 0 .000925925 .000925925 .00000001 .001851851)
V3B 39 38 PULSE(0 214 0 .000925925 .000925925 .00000001 .001851851)
V4B 42 41 PULSE(0 214 0 .000925925 .000925925 .00000001 .001851851)
V5B 45 44 PULSE(0 214 0 .000925925 .000925925 .00000001 .001851851)
V6B 48 47 PULSE(0 214 0 .000925925 .000925925 .00000001 .001851851)
*
S1X 31 32 33 0 RELAYX
S2X 34 35 36 0 RELAYX
S3X 37 38 39 0 RELAYX
S4X 40 41 42 0 RELAYX
S5X 43 44 45 0 RELAYX
S6X 46 47 48 0 RELAYX
*
.MODEL RELAYX VSWITCH(RON=1E-3 ROFF=1E6 VON=100.4 VOFF=100.1)
*
.TRAN .000006 2
*
.PROBE
.END
```

3) Once the simulation is run, Probe can be used to display traces of voltages and currents throughout the circuit.

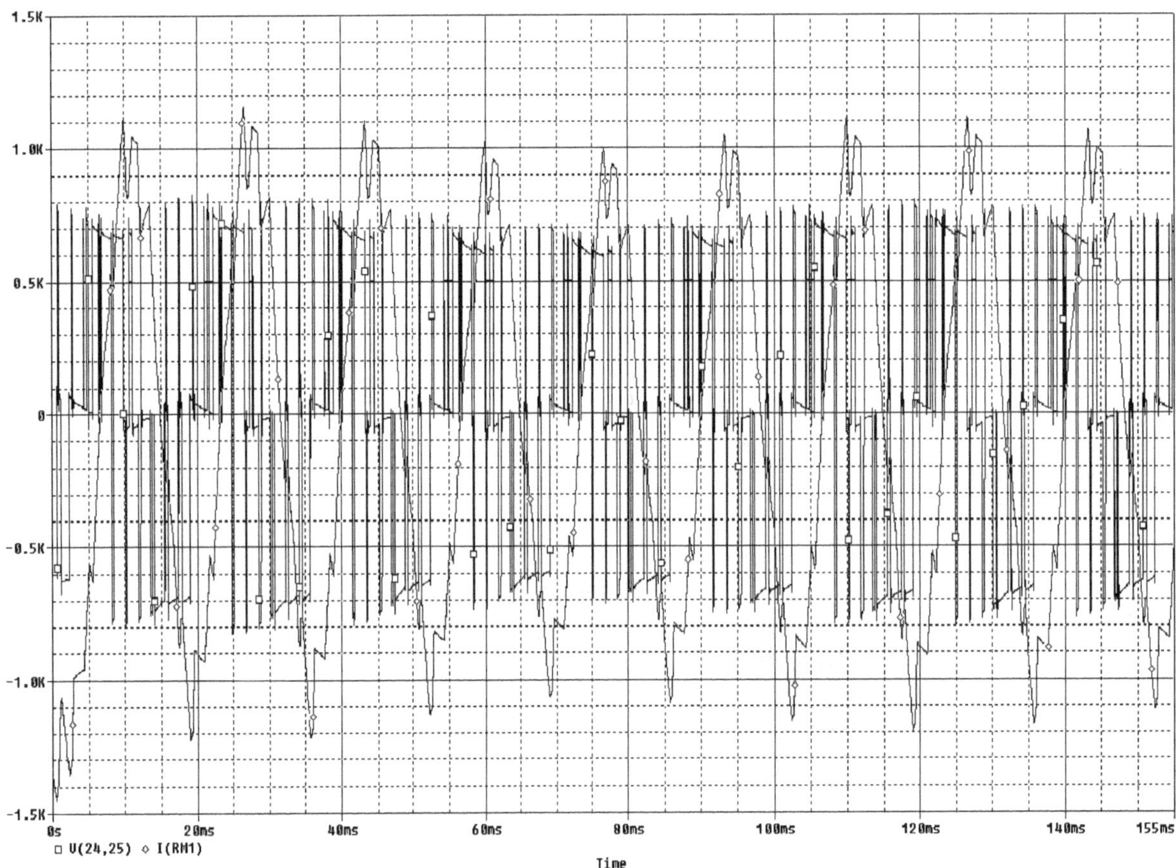

Figure 4-3-1-3 Line-to-line voltage applied to the motors and current through one phase of the motors. Probe "Zoom Area" has been used to display this region of the output. C equals .187 F.

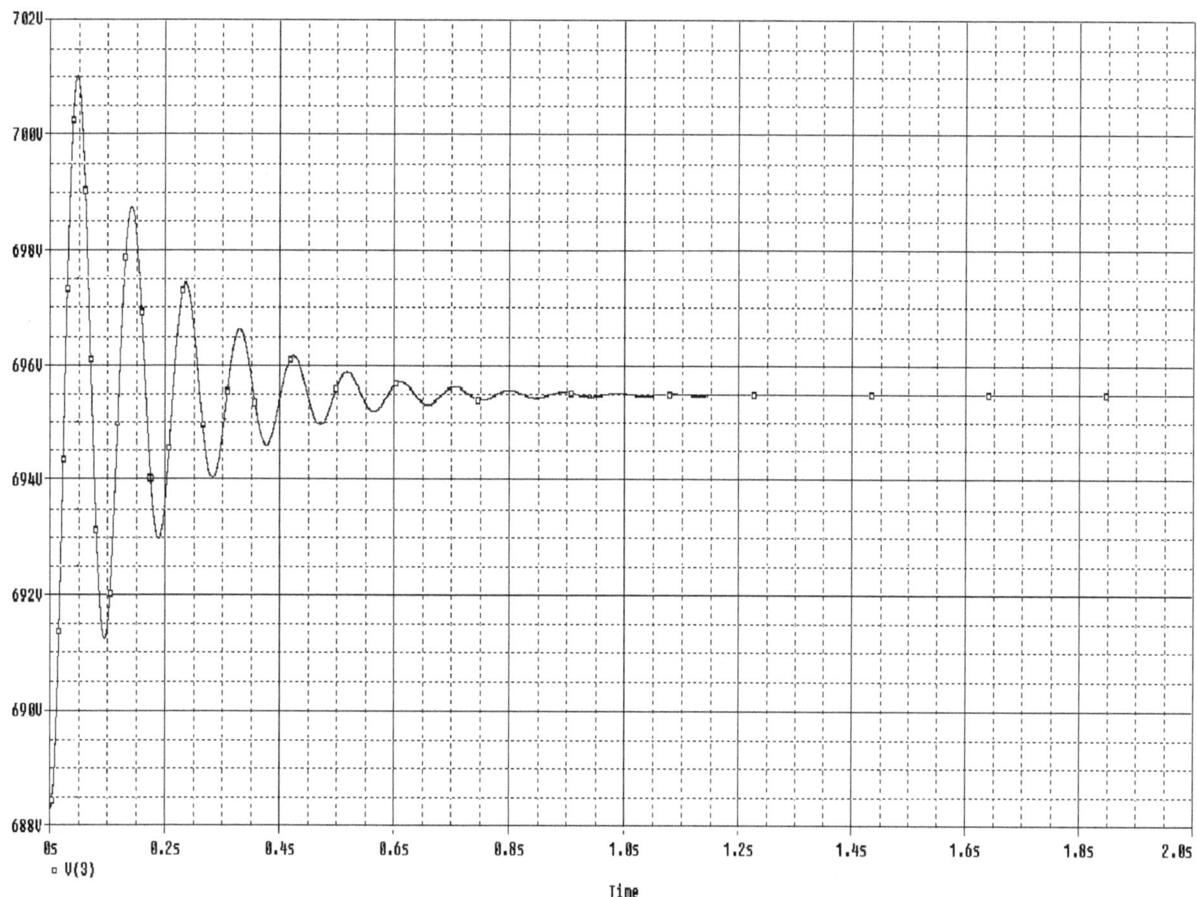

Figure 4-3-1-4 Voltage at the input to the inverter. C equals .187 F.

4) Figure 4-3-1-4 shows that the noise produced by the inverter at its input is much less than that from the similarly constructed six-step inverter of Figure 4-2-5. Probe is used to look at a blowup of the waveform in the region from .85 to 1.087 seconds. See Figure 4-3-1-5.

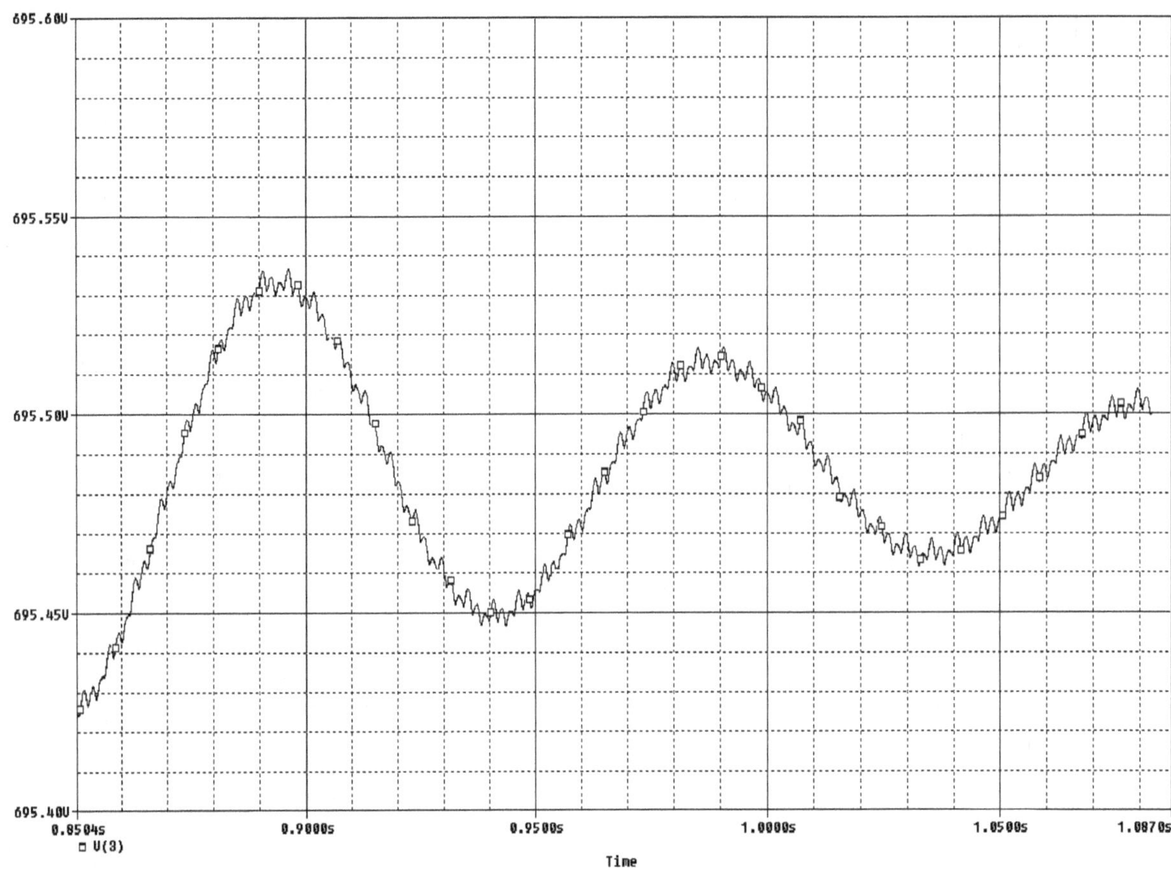

Figure 4-3-1-5 Voltage at the input to the inverter in the region from .85 to 1.087 seconds of analysis. Probe "Zoom Area" has been used to display this region of the output. C equals .187 F.

4.3.2 PWM INVERTER WITH TIMED-PULSE CONTROL

PSpice can be used to represent a timed-pulse PWM inverter. The pulses are simply entered one at a time for each pole of the inverter. The process of entering pulse data may be tedious, if many pulses are used per waveform period.

Here 3 pulses are used per half period. This is within the capabilities of the 64 node limit of the PSpice A/D demo. If an analysis with more 5 pulses per half period was needed, it would be necessary to use the unlimited node capability of a purchased copy of PSpice.

The schematic for a PWM inverter with a timed-pulse control is the same as that for the six-step inverter except that a timed-pulse control is used rather than a six-step control. Replace the six-step switch control circuit of Figure 4-2-2 with the timed-pulse PWM switch control circuit of Figure 4-3-2-1 to create the schematic.

Only use C equal to .187 F in this analysis.

Note that the netlist uses the ".PARAM" statement to input frequency, F. This makes PSpice carry out many needed calculations that would otherwise have to be done manually. To modify the frequency the programmer only needs to change the value of F once in the ".PARAM" statement.

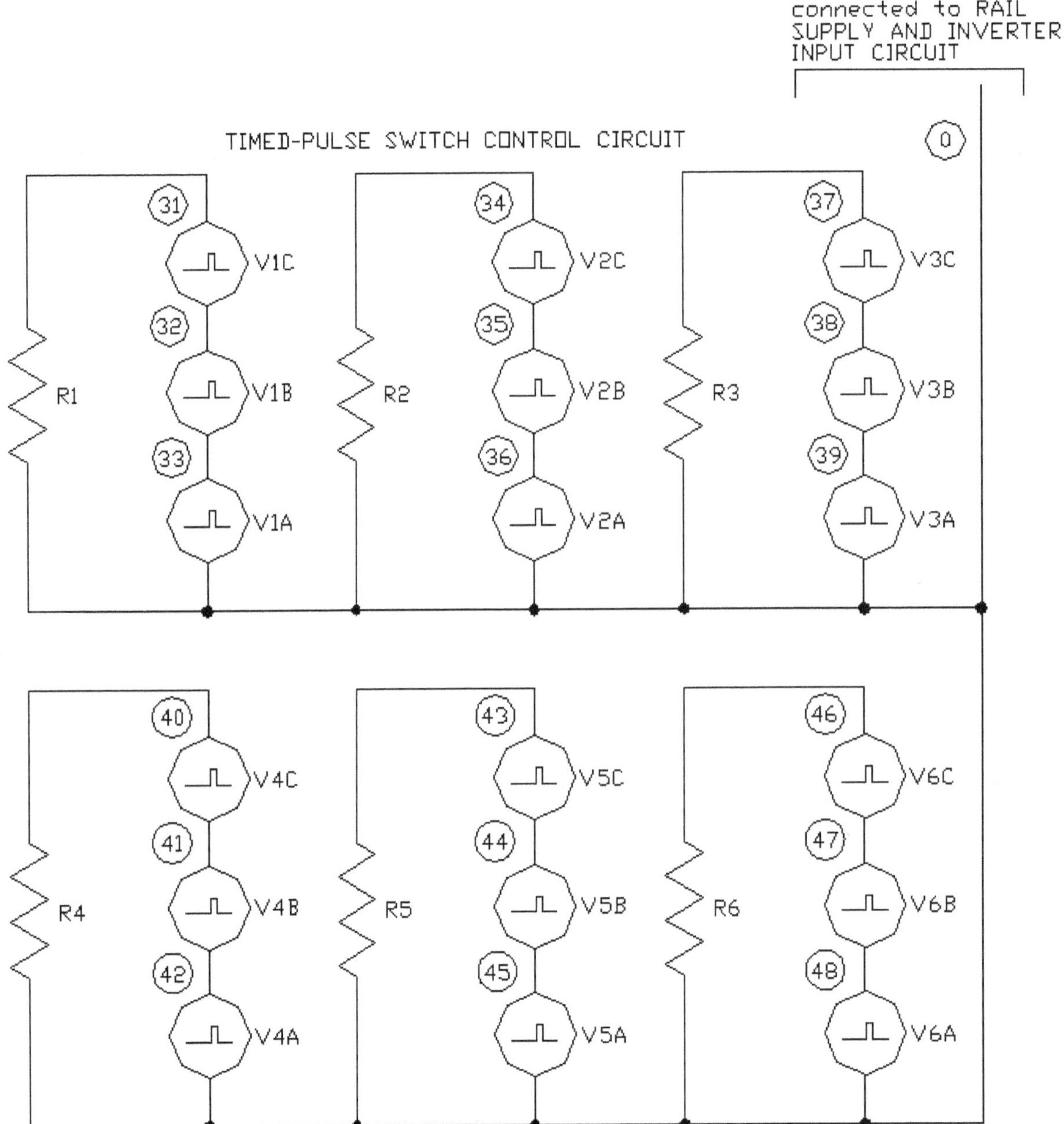

Figure 4-3-2-1 Control circuit of a timed-pulse controlled PWM IGBT inverter receiving power from a third rail and supplying power to two railcar traction motors. The rest of the schematic is the same as seen in Figure 4-2-2.

When the resistors of Figure 4-3-2-1 receive positive voltage their corresponding switches (the ones representing IGBTs) close. In Figure 4-3-1-2 the pulses are shown for resistors R1 and R4 for the A phase. R1 will receive positive voltage pulses and close its corresponding switch S1 when V1A, V1B, or V1C is 1 volt. R4 will receive positive voltage and close its corresponding switch S4 when V4A, V4B, or V4C is 1 volt.

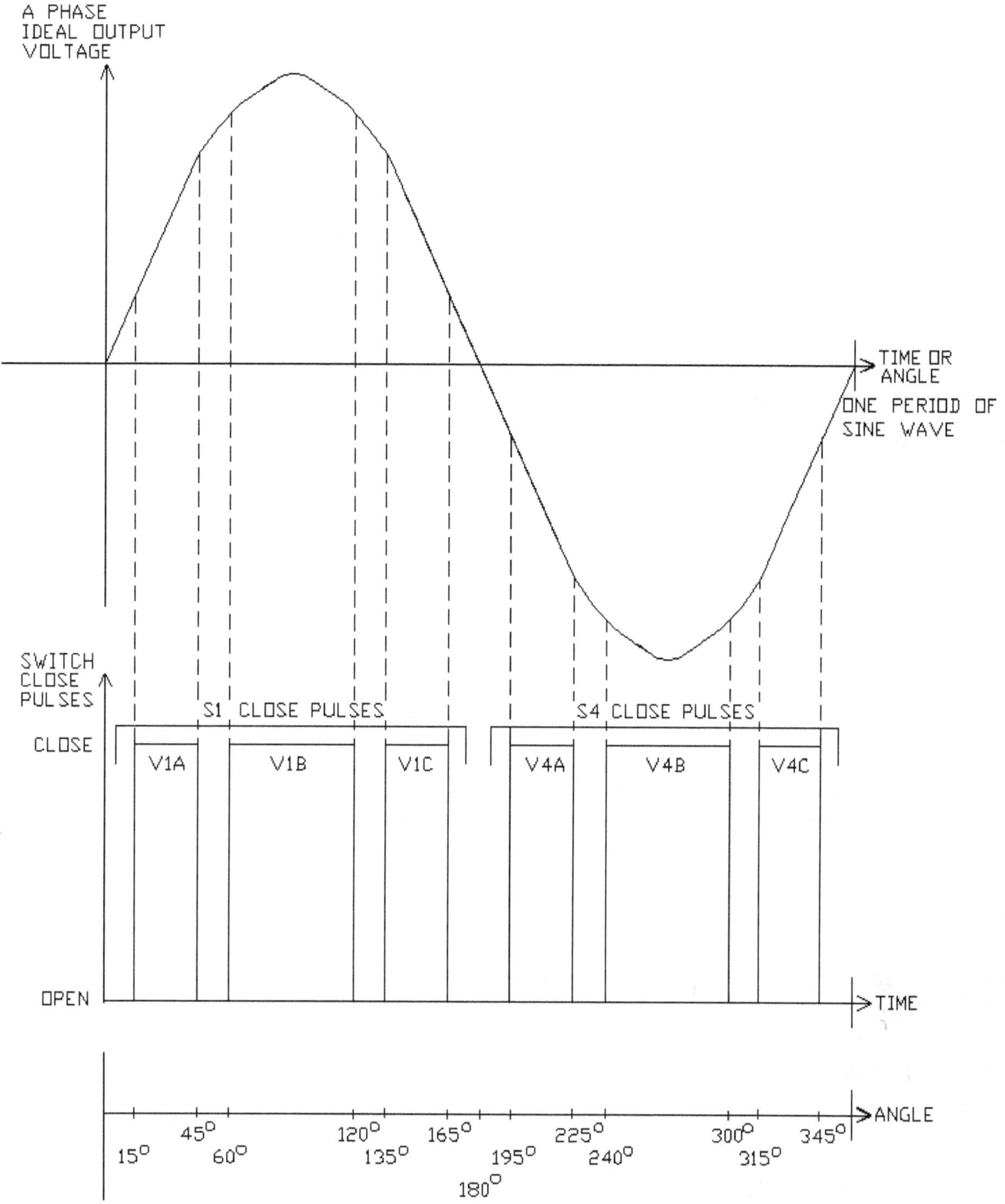

Figure 4-3-2-2 Timed pulses for controlling the A phase PWM voltage. The upper graph shows the ideal sine wave that the pulses are based on.

PROCEDURE:

1) Values for the supply voltage and components are the same as those in Section 4.2 except as noted in the following:

DELETE THE FOLLOWING:

Description	**Symbol**	**Value and units**
Switch turn-on voltage pulses	V1, V2, V3A, V3B, V4, V5A, V5B, V6	.0166667 second period 1 V pulses of varying widths to turn on the switches. These are shown in Figure 4-2-3.

ADD THE FOLLOWING:

Description	**Symbol**	**Value and units**
Switch turn-on voltage pulses	V1A, V1B, V1C, V2A, V2B, V2C, V3A, V3B, V3C, V4A, V4B, V4C, V5A, V5B, V5C, V6A, V6B, V6C	.0166667 second period 1 V pulses of varying widths to turn on the switches that represent IGBTs. The pulses for the A phase are shown in Figure 4-3-2-2.

2) Write the netlist program:

```
Figure 4-3-1b.CIR PWM CONTROLLED BY TIMED PULSES
* Plotting motor input current and voltage and voltage at the inverter
* input versus time.
*
V 1 0 700
*
RR 1 2 .00863
LR 2 3 1E-9
*
CF 3 4 .001E-6
LF 3 4 .0004
*
L 4 5 .0008
C 0 5 .187
*
DF1 26 5 TYPE1
DF2 25 5 TYPE1
DF3 24 5 TYPE1
```

```
DF4 0 26 TYPE1
DF5 0 25 TYPE1
DF6 0 24 TYPE1
*
DG1 6 26 TYPE1
DG2 9 25 TYPE1
DG3 12 24 TYPE1
DG4 15 0 TYPE1
DG5 18 0 TYPE1
DG6 21 0 TYPE1
*
DS1 8 26 TYPE1
DS2 11 25 TYPE1
DS3 14 24 TYPE1
DS4 17 0 TYPE1
DS5 20 0 TYPE1
DS6 23 0 TYPE1
*
.MODEL TYPE1 D(RS=.03 BV=10000 IBV=50m)
*
CS1 7 8 .000006
CS2 10 11 .000006
CS3 13 14 .000006
CS4 16 17 .000006
CS5 19 20 .000006
CS6 22 23 .000006
*
LS1 5 7 .8E-6
LS2 5 10 .8E-6
LS3 5 13 .8E-6
LS4 26 16 .8E-6
LS5 25 19 .8E-6
LS6 24 22 .8E-6
*
RS1 8 26 .1
RS2 11 25 .1
RS3 14 24 .1
RS4 17 0 .1
RS5 20 0 .1
RS6 23 0 .1
```

```
*
RM1  24  27  .222
RM2  26  28  .222
RM3  25  29  .222
*
LM1  30  27  .000589
LM2  30  28  .000589
LM3  30  29  .000589
*
R1  31  0  1000
R2  34  0  1000
R3  37  0  1000
R4  40  0  1000
R5  43  0  1000
R6  46  0  1000
*
S1  5  6  31  0  RELAY
S2  5  9  34  0  RELAY
S3  5  12  37  0  RELAY
S4  26  15 40  0  RELAY
S5  25  18  43  0  RELAY
S6  24  21  46  0  RELAY
*
.MODEL RELAY VSWITCH(RON=1E-3 ROFF=1E6 VON=.6 VOFF=.4)
*
.PARAM F = 60;  Fundamental frequency received by motor
*
V1A  33  0  PULSE(0  100V  {15/360/F }  0  0  {30/360/F}  {1/F})
V1B  32  33  PULSE(0  100V  {60/360/F }  0  0  {60/360/F}  {1/F})
V1C  31  32  PULSE(0  100V  {135/360/F}  0  0  {30/360/F}  {1/F})
V4A  42  0  PULSE(0  100V  {15/360/F  + 180/360/F}  0  0  {30/360/F}  {1/F})
V4B  41  42  PULSE(0  100V  {60/360/F  + 180/360/F}  0  0  {60/360/F}  {1/F})
V4C  40  41  PULSE(0  100V  {135/360/F + 180/360/F}  0  0  {30/360/F}  {1/F})
V2A  36  0  PULSE(0  100V  {15/360/F  + 120/360/F}  0  0  {30/360/F}  {1/F})
V2B  35  36  PULSE(0  100V  {60/360/F  + 120/360/F}  0  0  {60/360/F}  {1/F})
V2C  34  35  PULSE(0  100V  {135/360/F + 120/360/F}  0  0  {30/360/F}  {1/F})
V5A  45  0  PULSE(0  100V  {15/360/F  + 180/360/F + 120/360/F}  0  0  {30/360/F}  {1/F})
V5B  44  45  PULSE(0  100V  {60/360/F  + 180/360/F + 120/360/F}  0  0  {60/360/F}  {1/F})
V5C  43  44  PULSE(0  100V  {135/360/F + 180/360/F + 120/360/F}  0  0  {30/360/F}  {1/F})
V3A  39  0  PULSE(0  100V  {15/360/F  + 240/360/F}  0  0  {30/360/F}  {1/F})
```

```
V3B 38 39 PULSE(0 100V {60/360/F + 240/360/F} 0 0 {60/360/F} {1/F})
V3C 37 38 PULSE(0 100V {135/360/F + 240/360/F} 0 0 {30/360/F} {1/F})
V6A 48 0 PULSE(0 100V {15/360/F + 180/360/F + 240/360/F} 0 0 {30/360/F} {1/F})
V6B 47 48 PULSE(0 100V {60/360/F + 180/360/F + 240/360/F} 0 0 {60/360/F} {1/F})
V6C 46 47 PULSE(0 100V {135/360/F + 180/360/F + 240/360/F} 0 0 {30/360/F} {1/F})
*
.TRAN .000006 2
*
.PROBE
.END
```

3) Once the simulation is run, Probe can be used to display traces of voltages and currents throughout the circuit.

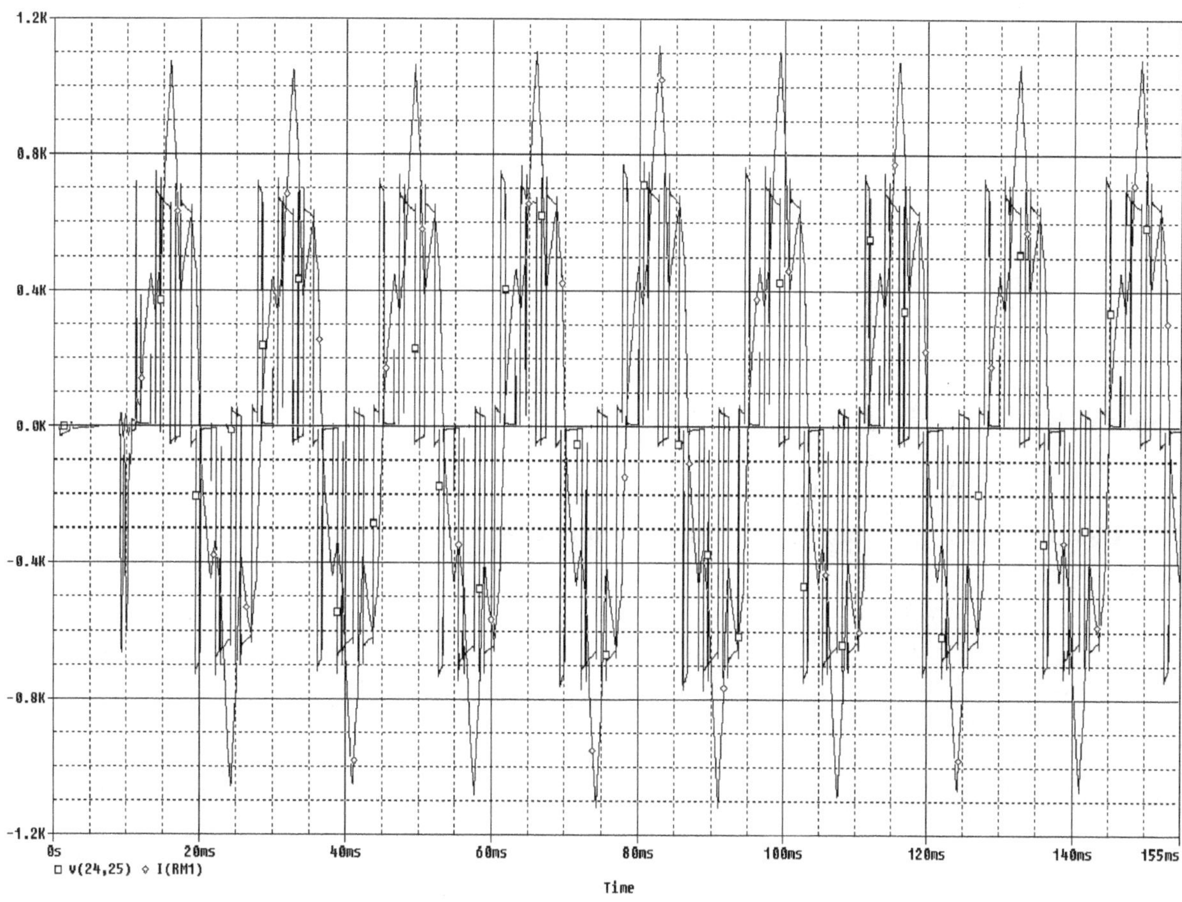

Figure 4-3-2-3 One line-to-line voltage applied to the motors and the current through one phase of the motors. Probe "Zoom Area" has been used to display this region of the output. C equals .187 F.

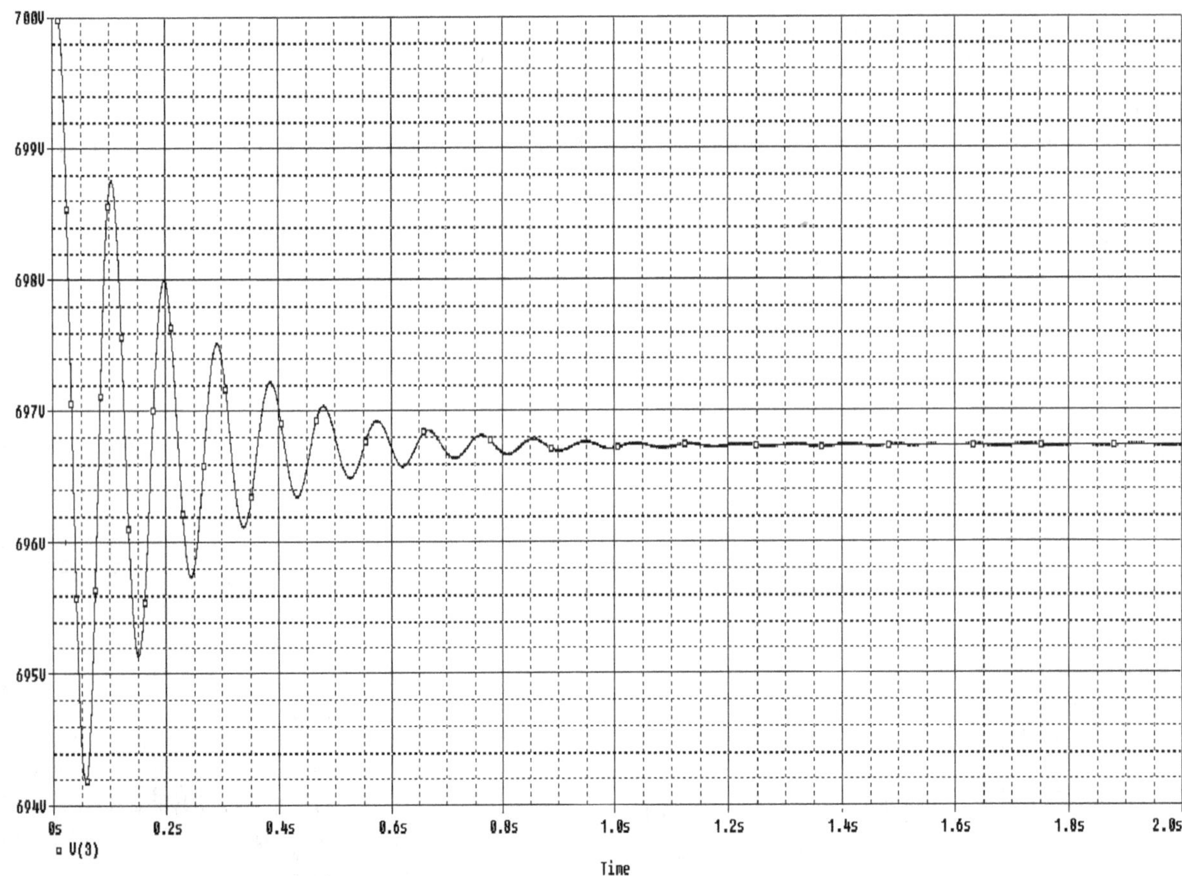

Figure 4-3-2-4 Voltage at the input to the inverter. C equals .187 F.

4) Figure 4-3-2-4 shows that the noise produced by the inverter at its input is less than that from the six-step inverter of Figure 4-2-5, although more than that of the triangle/sine wave method of Figure 4-3-1-4.

4.4 DC TO DC CHOPPER

PROBLEM: A DC chopper inputs 700 V_{DC} and outputs 350 V_{DC}. The output supplies a 12.1 Ω resistive load. The switching frequency is 240 Hz. Switching is done with an IGBT. A large power inductor, .04 H and .25 Ω DC, is used to smooth the output voltage. What will be the output voltage and power waveforms? How much waste heat is produced by the inductor and snubber? What would be the effect of using two of the power inductors in series rather than one?

DISCUSSION:

DC choppers work by periodically connecting a higher DC input voltage supply to their lower voltage output. They have inductors and/or capacitors to assure that their output voltage will not instantly increase to the input voltage level. Usually, choppers have a fixed switching period. During that switching period, voltage sensing circuitry at the chopper's output directs the IGBT (or other power electronic switching device) to connect the high voltage input to the output circuit for a percent of the switching period. For the greatest loads, the switch might be on almost 100% of the pulse time. For no-load, the switch would usually be off.

In this, simplified no feedback PSpice simulation, the user manually inserts different IGBT percent on-time values into the netlist until the desired average output voltage results.

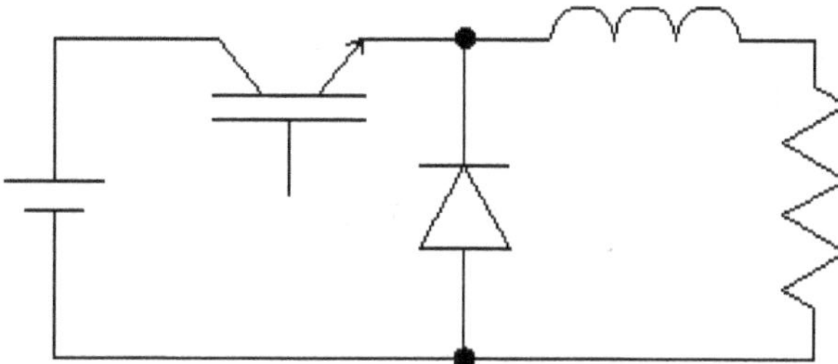

Figure 4-4-1 Basic power circuit of a DC to DC IGBT chopper.

PROCEDURE:

1) Draw the power circuit.

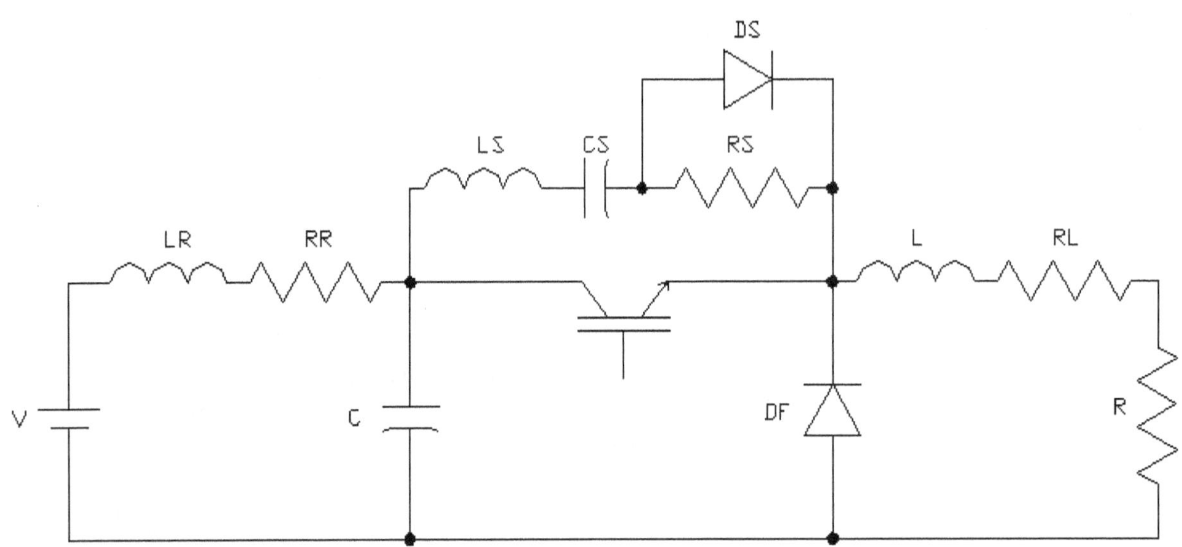

Figure 4-4-2 Chopper power circuit with the addition of input impedances, snubber circuit, smoothing inductor, and load inductance.

2) Redraw the circuit with node numbers, a switch/diode circuit in place of the IGBT, and a timing circuit.

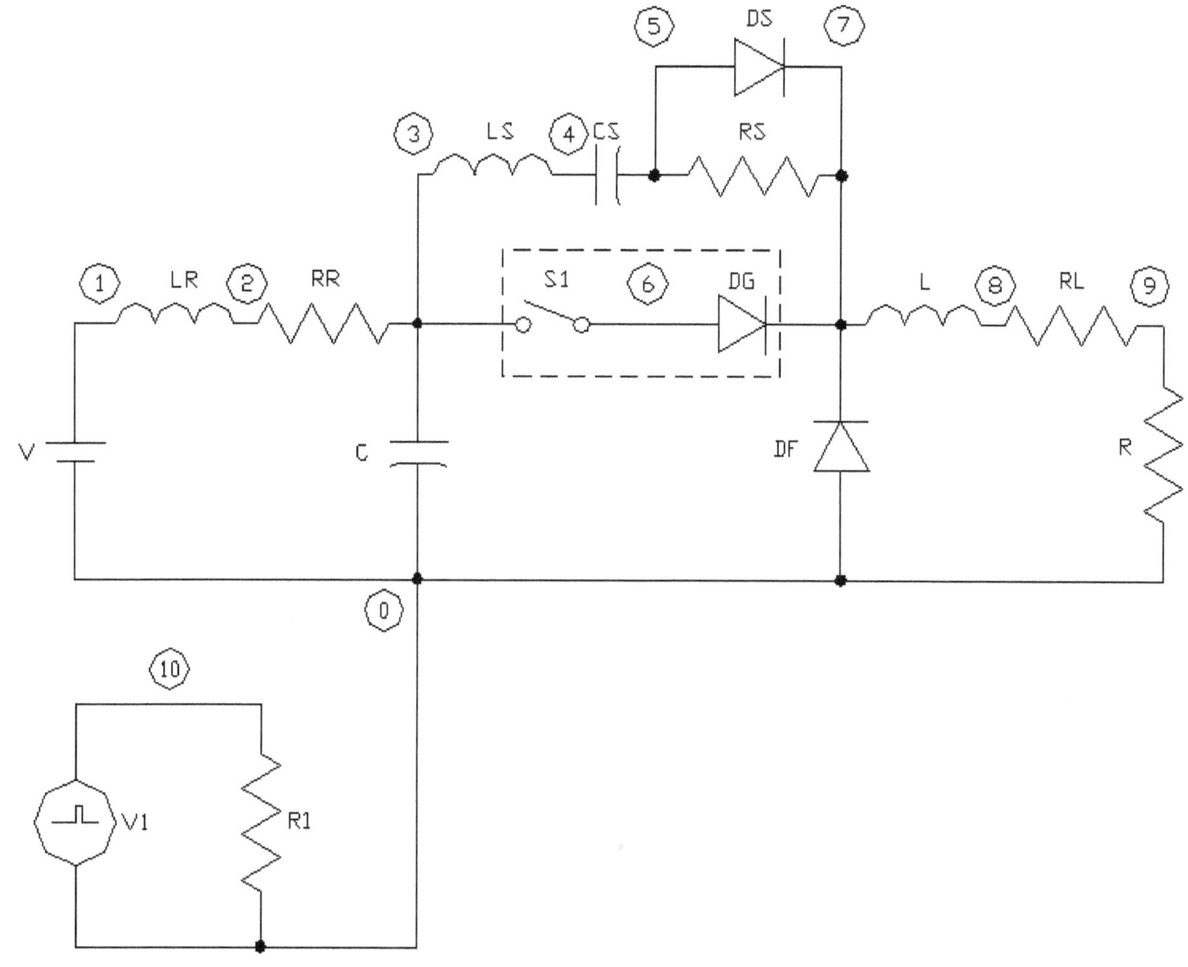

Figure 4-4-3 Chopper circuit ready for analysis with PSpice.

3) Values for the supply voltage and components are:

Description	Symbol	Value and units
Switching frequency	F	240 Hz
Percent S conducts	P	55 %
Supply voltage	V	700 V_{DC}
Rail inductance	LR	1E-9 H
Rail resistance	RR	.00863 Ω
Charging capacitor capacitance	C	.187 F

Continued

Effective snubber circuit inductance	LS	.8E-6 H
Capacitance	CS	.000006 F
Snubber resistor resistances	RS	.1 Ω
Diodes	DF, DG, and DS	Effective on series resistance = .03 Ω, Breakdown Voltage = 10000 V_{DC}, Current at Breakdown Voltage = 50 milliamps
Inductance of smoothing inductor	L	.04 H
Effective resistance of smoothing inductor	RL	.25 Ω DC resistance plus an approximation of magnetic loss effective resistance, .25*F/60
Load resistance	R	12.1 Ω
Switch turn-on voltage pulses	V1	1/F second period 100 V pulses of varying widths to turn on the switch.
Switch circuit resistor resistance	R1	1000 Ω
Switch	S1	On resistance = 1E-3 Ω, Off resistance = 1E6 Ω, Turn-on voltage = .6 V, Turn-off voltage = .4 V

4) Write the netlist program:

```
Figure 4-4-3.CIR DC TO DC CHOPPER
* Plotting load voltage and inductor power loss versus time for two different
* smoothing inductor inductances.
*
.PARAM F = 240;  Switching frequency, Hz
*
.PARAM P = 55;  Percent of period switch S conducts
*
V  1  0  700
*
LR  1  2  1E-9
RR  2  3  .00863
*
C  3 0  .187
```

```
*
LS 3 4 .8E-6
CS 4 5 .000006
RS 5 7 .1
*
DS 5 7 TYPE1
DF 0 7 TYPE1
DG 6 7 TYPE1
.MODEL TYPE1 D(RS=.03 BV=10000 IBV=50m)
*
L 7 8 .04
* Note L is doubled to .08 to model two smoothing inductors in series.
*
RL 8 9 {.25 + .25*F/60}
* Note RL is composed of a DC component and an AC frequency dependent component.
* Note RL is doubled to {.5 + .5*F/60} to model two smoothing inductors in series.
*
R 9 0 12.1
*
R1 10 0 1000
*
S1 3 6 10 0 RELAY
*
.MODEL RELAY VSWITCH(RON=1E-3 ROFF=1E6 VON=.6 VOFF=.4)
*
V1 10 0 PULSE(0 100V 0 0 0 {P/100/F} {1/F})
*
.TRAN .000006 .06
*
.PROBE
.END
```

5) Once the simulation is run, Probe can be used to display traces of output voltage and inductor power usage.

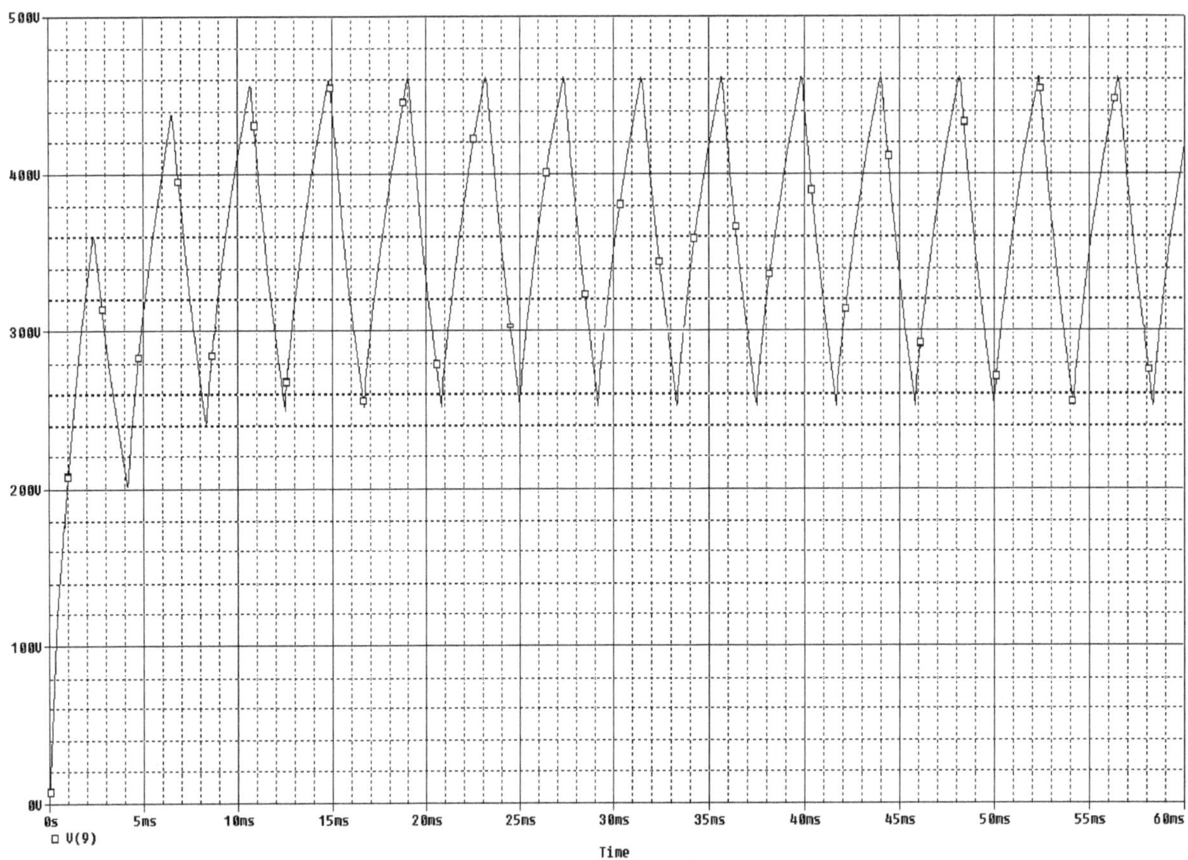

Figure 4-4-4 Voltage applied to the load resistance with one smoothing inductor.

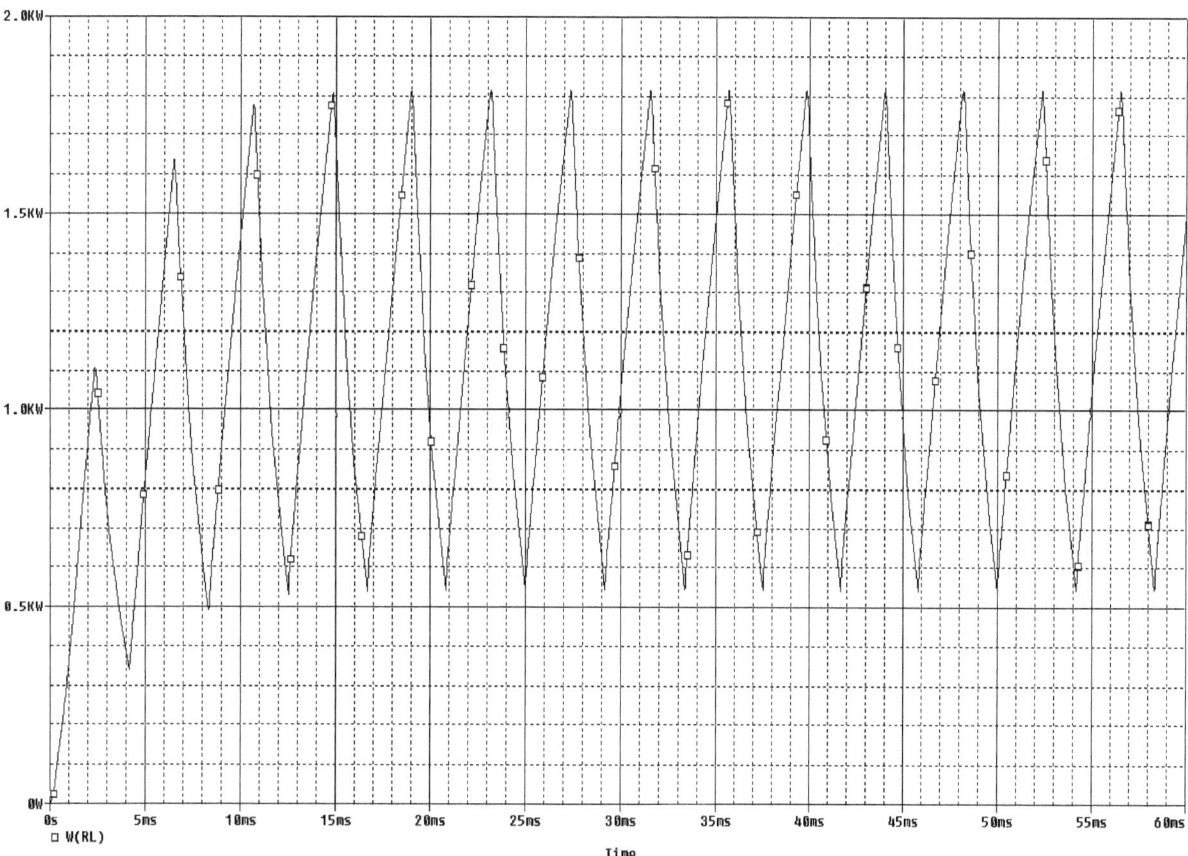

Figure 4-4-5 Power used by the one smoothing inductor.

6) Double the values of L and RL and rerun PSpice. Leave the percent on time, P, the same for this.

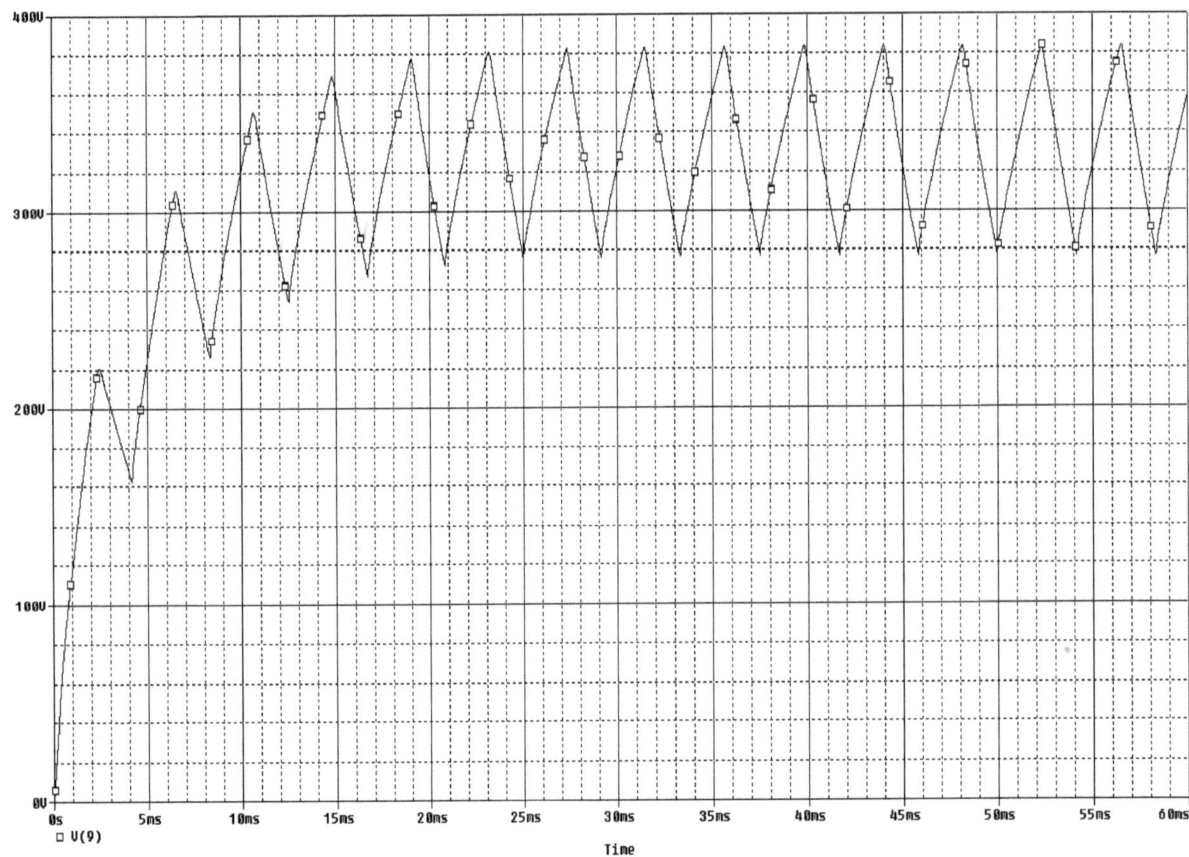

Figure 4-4-6 Voltage applied to the load resistance with two smoothing inductors in series.

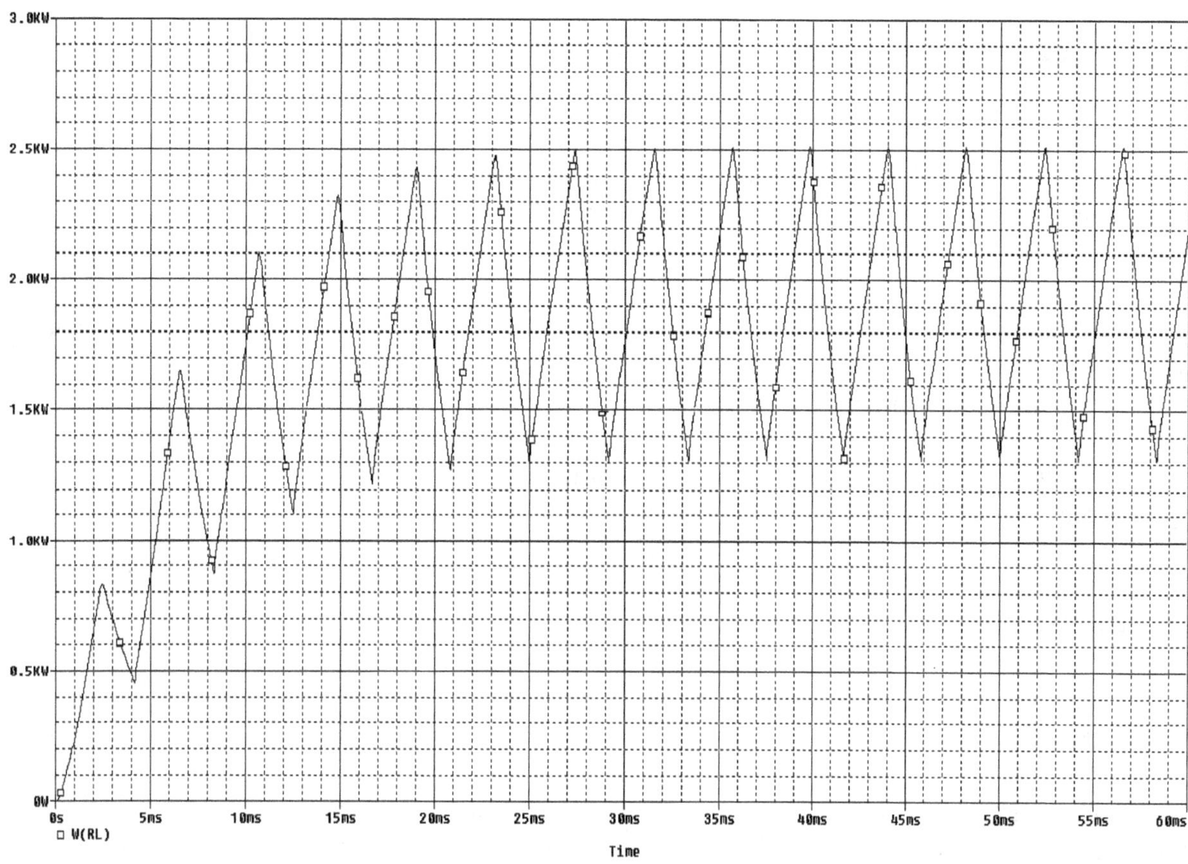

Figure 4-4-7 Power used by two power smoothing inductors in series.

5.0 POWER EXAMPLE PROBLEMS

Power circuits can be difficult to solve due to their size and their use of three-phase. Before computers were available, only balanced three-phase faults could be easily solved. Unbalanced faults either were assumed to be less serious than balanced faults or solved manually by the symmetrical components method.

Programs such as ETAP and SKM are designed for power circuit analysis. PSpice is not designed for it, but can do it. The flexibility of PSpice makes it better able to solve unusual power circuits than the ETAP and SKM programs.

All the examples in this section use induction motors represented as resistances in series with inductances. This was discussed at the beginning of Section 4.2.

5.1 TRANSFORMERS

Power circuits often use transformers. Unfortunately, transformers make PSpice analysis more difficult.

There are several ways of analyzing transformer circuits with PSpice. One way is to create PSpice transformer equivalent circuits using inductors with mutual inductances. Another way is to eliminate the transformers from the circuits and reflect (convert) the circuits' impedances, voltages, and currents on one side or the transformer to the values they would have on the other side of the transformer circuit.

Each method has strengths and limitations. The mutually induced inductors PSpice equivalent transformer circuit works well with loaded transformers, but cannot be used with unloaded transformers. Reflected value equivalent transformer circuits work well with single-phase circuits and Y-Y three-phase circuits. Δ-Δ three-phase transformer circuits can be analyzed by converting them to Y-Y. However, the reflected value transformer equivalent circuits cannot predict the phase shift that occurs with three-phase Y-Δ and Δ-Y transformer circuits.

The symmetrical component method uses reflected value transformer equivalent circuits to represent three-phase transformer circuits. Then those reflected value circuits are incorporated in symmetrical component phase circuits. However, the symmetrical component method requires significant training and is tedious.

Only the PSpice transformer equivalent circuit and reflected value methods will be used here.

5.1.1 PSPICE TRANSFORMER EQUIVALENT CIRCUIT METHOD

Look at the usual single-phase transformer equivalent circuit:

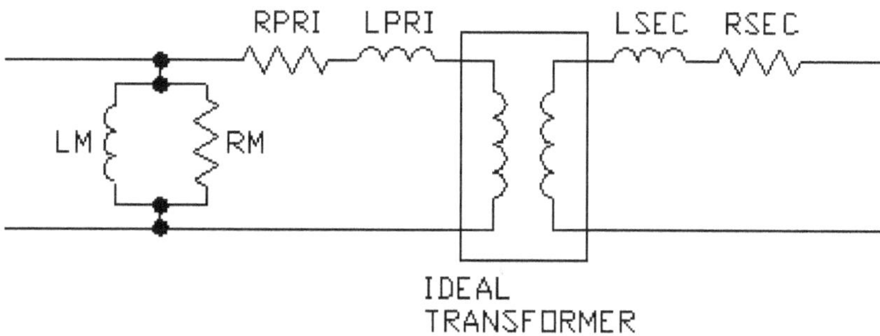

Figure 5-1-1-1 Usual single-phase transformer equivalent circuit, not directly solvable by PSpice.

For PSpice to simulate a transformer the commonly used equivalent circuit must be modified. First, a current path must be made between the primary and secondary windings. PSpice requires a current path between all circuit components. A very high resistance can be used as the current path, so its effect on circuit operation is negligible. Second, the values of the special PSpice mutual inductances, LPRIM and LSECM, need to be chosen. They must follow the equation LPRIM = LSECM x [(Secondary Turns)/(Primary Turns)]^2 .and the values of LPRIM and LSECM need to have their inductive impedances, at the transformed frequency, at least ten times greater than the impedances of the components in series with them. This may require large inductances. Keep in mind that LPRIM and LSECM are not real inductances, but are just values needed to make PSpice simulate a transformer.

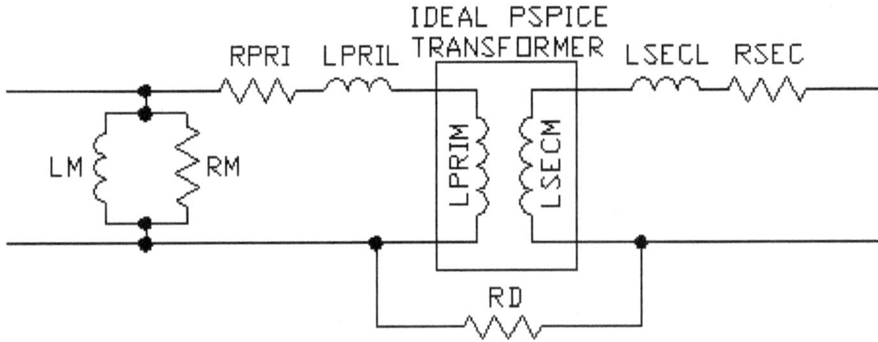

Figure 5-1-1-2 PSpice transformer equivalent circuit.

In PSpice netlists the statement that accomplishes the transformer simulation is:

KPRI-SEC LPRIM LSECM 1.00

5.1.2 TRANSFORMER REFLECTED VALUE METHOD

The primary side values are reflected (converted) to the secondary side by the equations:

Reflected primary to secondary side impedance value = (primary side impedance value)x[(secondary turns)/(primary turns)]^2

Reflected primary to secondary side voltage value = (primary side voltage value)x[(secondary turns)/(primary turns)]

Reflected primary to secondary side current value = (primary side current value)x[(primary turns)/(secondary turns)]

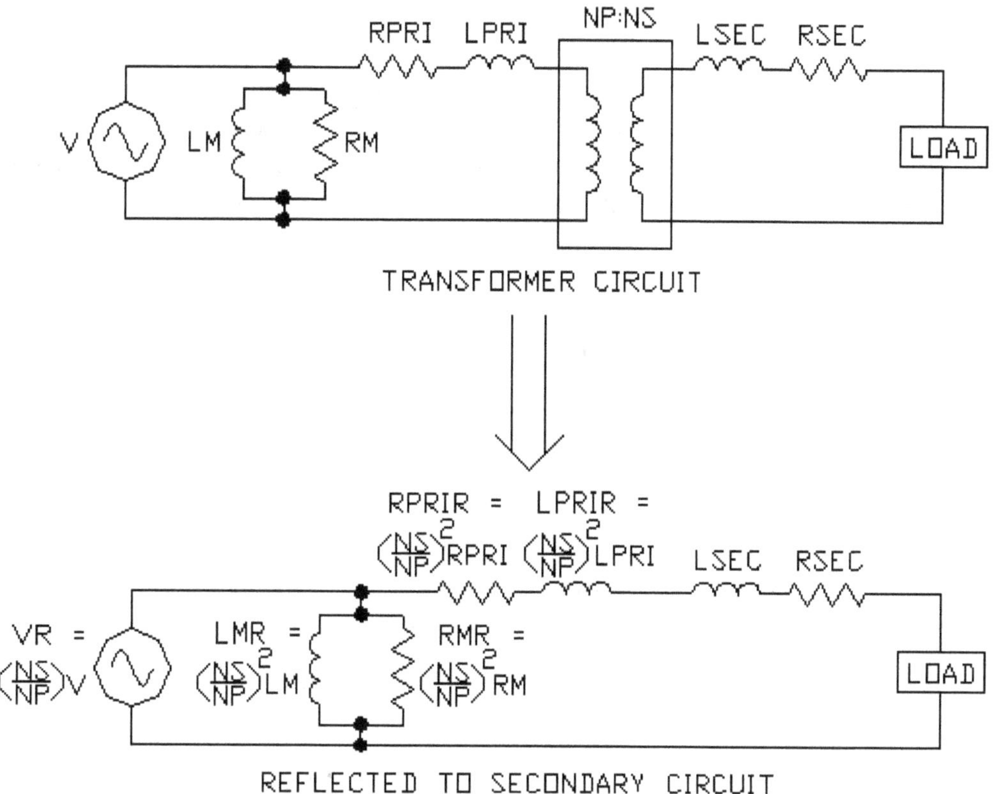

Figure 5-1-2-1 Conversion of a transformer circuit to an equivalent reflected value circuit.

After the reflected circuit is solved, voltage and current values found in the reflected part of the circuit need to be un-reflected to get actual values.

5.2 STEADY-STATE BALANCED THREE-PHASE LOAD FLOW

PROBLEM: Two induction motors and an electronic load are supplied three-phase through a three-phase transformer. The voltage supply and loads are all balanced. What are the steady-state line-to-line voltages and line currents at the transformers and loads? Solve it with the PSpice transformer equivalent circuit method and then with the reflected values method.

5.2.1 SOLVED BY PSPICE TRANSFORMER EQUIVALENT CIRCUIT METHOD

PROCEDURE:

1) Draw the basic one-line diagram.

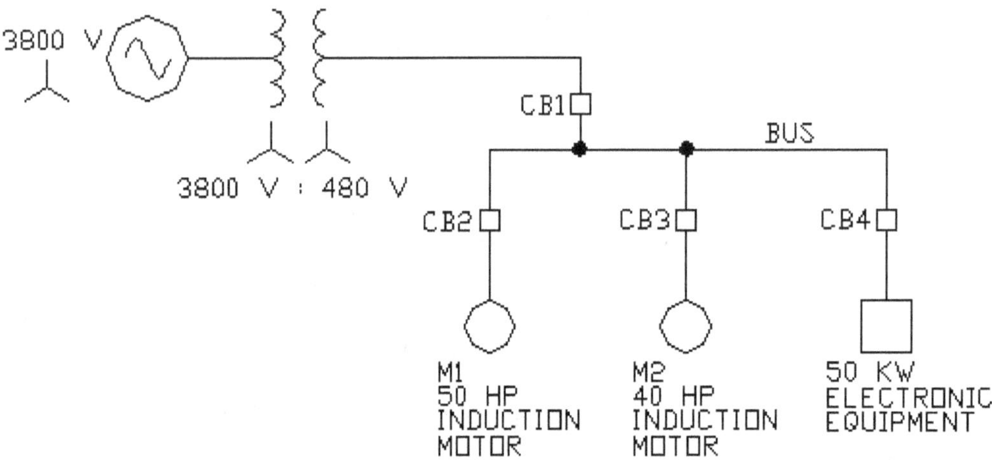

Figure 5-2-1-1 One-line diagram, steady-state load flow.

2) Since the source and loads are balanced, it is only necessary to solve one line-to-neutral portion of the three-phase circuit. Only the A phase is drawn here. It is drawn with equivalent line impedances, a PSpice transformer equivalent, motor and electronic equipment equivalent impedances, and node numbers.

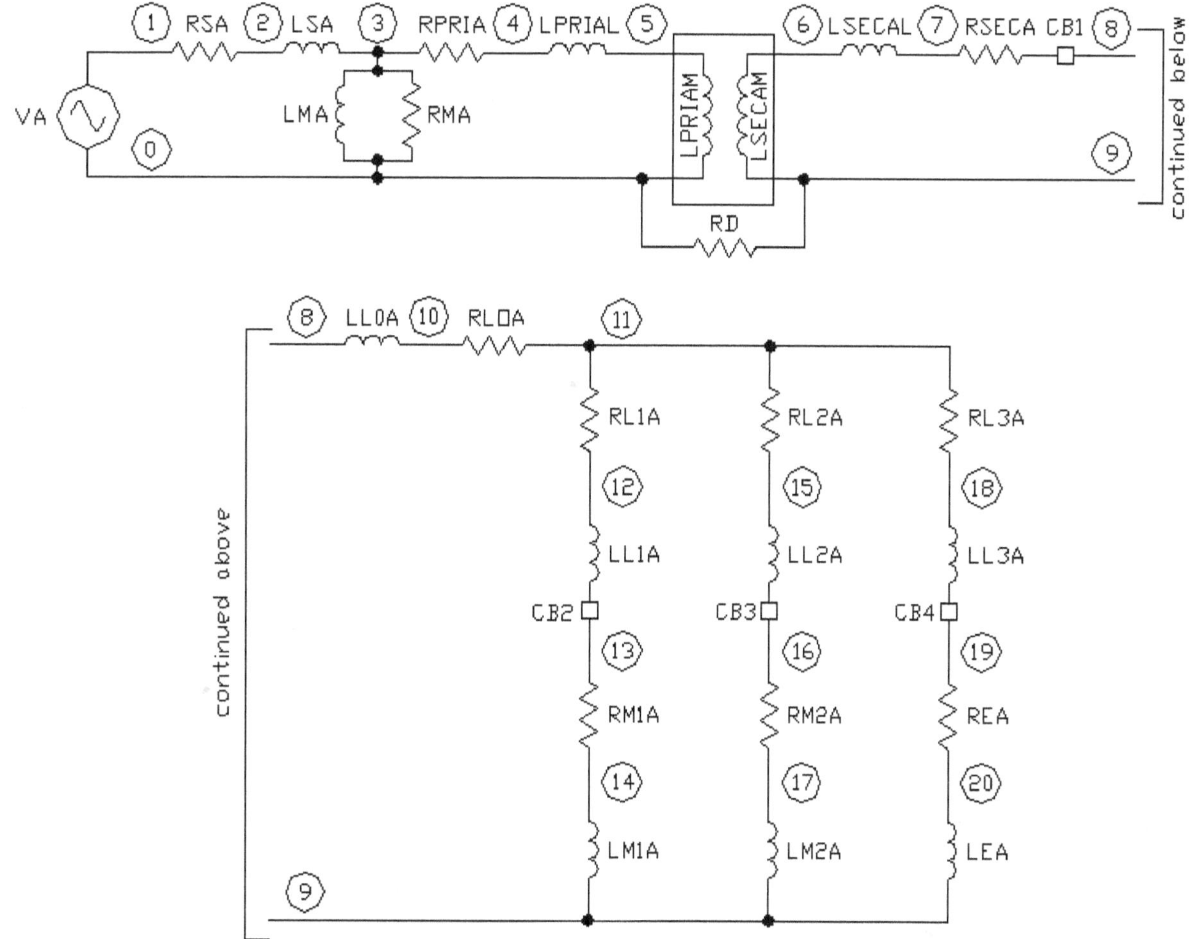

Figure 5-2-1-2 Line-to-neutral A phase equivalent circuit with a PSpice transformer equivalent circuit. The circuit is ready for analysis by PSpice.

3) Values for the supply voltage and components are:

Description	Symbol	Value and units
A phase line-to-neutral AC input sinusoidal peak voltage	VA	3103 V_{ACpeak}, 60 Hz frequency, 0° phase angle
System equivalent line resistance	RSA	.2 Ω

Continued

System equivalent line inductance	LSA	.0008 H
Transformer magnetizing inductance per phase	LMA	25 H
Transformer magnetizing resistance per phase	RMA	22000 Ω
Transformer primary equivalent series resistance per line	RPRIA	.49 Ω
Transformer primary equivalent series leakage inductance per line	LPRIAL	.0011 H
Transformer primary equivalent series mutual inductance per line	LPRIAM	7.14 H (This was chosen so that 7.14x2πf = 7.14x377 = 2692 Ω >> RPRIA = .49 Ω)
Transformer secondary equivalent series mutual inductance per line	LSECAM	.114 H (This was chosen so that LSECAM = LPRIAMx$(480/3800)^2$ = .114 H and .114x2πf = .114x377 = 47 Ω >> RM2A = 4.3 Ω)
Transformer secondary equivalent series leakage inductance per line	LSECAL	.000018 H
Transformer secondary equivalent series resistance per line	RSECA	.0078 Ω
Transformer primary to secondary dummy resistance, needed to make PSpice work	RD	1E12 Ω
PSpice transformer factor that relates the primary voltage and current to the secondary voltage and current	KPRI-SECA	1.00
Feeder line inductance	LL0A	.00008 H
Feeder line resistance	RL0A	.003 Ω
Branch line resistances	RL1A, RL2A, RL3A	.005 Ω
Branch line inductances	LL1A, LL2A, LL3A	.0013 H
Motor M1 line-to-neutral resistance	RM1A	3.2 Ω
Motor M1 line-to-neutral inductance	LM1A	.0041 H

Continued

Motor M2 line-to-neutral resistance	RM2A	4.3 Ω
Motor M2 line-to-neutral inductance	LM2A	.0055 H
Electronic load line-to-neutral resistance	REA	3.9 Ω
Electronic load line-to-neutral impedance	LEA	.0055 H

4) Write the netlist program.

```
Figure 5-2-1-2.CIR LOAD FLOW WITH A PSPICE IDEAL TRANSFORMER EQUIVALENT
CIRCUIT
* Finding steady-state voltages and currents on a balanced three-phase circuit
*
VA 1 0 AC 3103 0
RSA 1 2 .2
LSA 2 3 .0008
*
LMA 3 0 25
RMA 3 0 22000
RPRIA 3 4 .49
LPRIAL 4 5 .0011
LPRIAM 5 0 7.14
LSECAM 6 9 .114
KPRI-SECA LPRIAM LSECAM 1.00
LSECAL 6 7 .000018
RSECA 7 8 .0078
RD 0 9 1E12
*
RL0A 8 10 .003
LL0A 10 11 .00008
*
RL1A 11 12 .005
LL1A 12 13 .0013
RM1A 13 14 3.2
LM1A 14 9 .0041
*
RL2A 11 15 .005
LL2A 15 16 .0013
RM2A 16 17 4.3
LM2A 17 9 .0055
```

```
*
RL3A  11  18  .005
LL3A  18  19  .0013
REA  19  20 3.9
LEA  20  9  .0055
*
.PRINT  AC  V(1,0)  I(RSA)  V(5,0)
*
.PRINT  AC  V(6,9)  I(RSECA)  V(11,9)
*
.PRINT  AC  V(13,9)  I(RL1A)
*
.PRINT  AC  V(16,9)  I(RL2A)
*
.PRINT  AC  V(19,9)  I(RL3A)
*
.AC  LIN  1  60  60
*
.END
```

5) Run the PSpice simulation. The AC data requested in the netlist is in the output file.:

```
**** 06/13/08 18:01:38 ******* PSpice Lite (August 2007) ****** ID# 10813 ****

Figure 5-2-1-2.CIR LOAD FLOW WITH A PSPICE IDEAL TRANSFORMER EQUIVALENT CIRCUIT

****     AC ANALYSIS                TEMPERATURE =   27.000 DEG C

******************************************************************************

 FREQ       V(1,0)      I(RSA)      V(5,0)

 6.000E+01   3.103E+03   3.349E+01   3.071E+03

 FREQ       V(6,9)      I(RSECA)    V(11,9)

 6.000E+01   3.880E+02   2.576E+02   3.805E+02

 FREQ       V(13,9)     I(RL1A)

 6.000E+01   3.561E+02   1.002E+02

 FREQ       V(16,9)     I(RL2A)

 6.000E+01   3.625E+02   7.594E+01

 FREQ       V(19,9)     I(RL3A)

 6.000E+01   3.598E+02   8.146E+01
```

Figure 5-2-1-3 Part of the output file from running the netlist of Figure 5-2-1-2.

5.2.2 SOLVED BY REFLECTED VALUES METHOD

1) Draw the one phase equivalent circuit again. But this time, reflect all of the values of the primary side of the transformer to the secondary side.

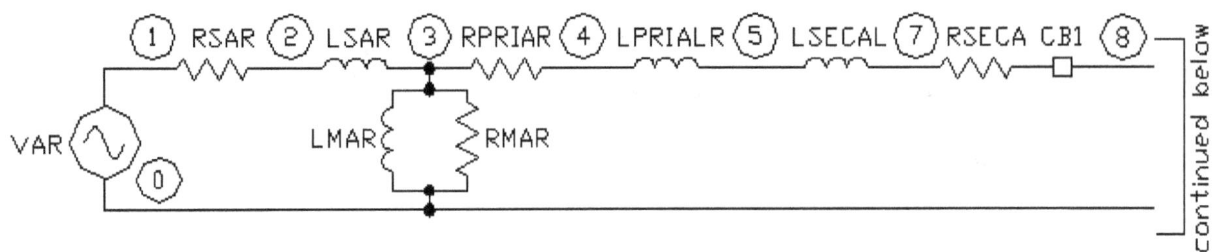

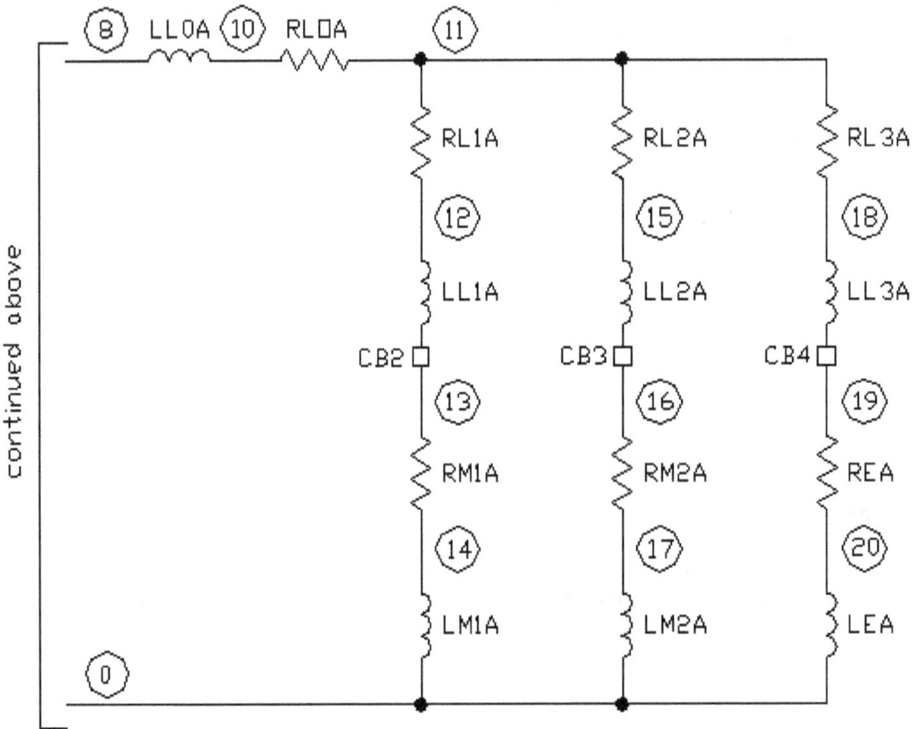

Figure 5-2-2-1 Line-to-neutral A phase equivalent circuit with values reflected from the primary (3800 V side) to the secondary (480 V side). The circuit is ready for analysis by PSpice.

2) Values for the supply voltage and components are:

Description	Symbol	Value and units
Reflected A phase line-to-neutral AC input sinusoidal peak voltage	VAR	3103x(480/3800) = 391.9 V_{ACpeak}, 60 Hz frequency, 0° phase angle
Reflected system equivalent line resistance	RSAR	.2 x$(480/3800)^2$ = .003191 Ω
Reflected system equivalent line inductance	LSAR	.0008 x$(480/3800)^2$ = .00001276 H
Reflected transformer magnetizing inductance per phase	LMAR	25 x$(480/3800)^2$ = .3989 H
Reflected transformer magnetizing resistance per phase	RMAR	22000 x$(480/3800)^2$ = 351.0 Ω
Reflected transformer primary equivalent series resistance per line	RPRIAR	.49 x$(480/3800)^2$ = .007818 Ω
Reflected transformer primary equivalent series leakage inductance per line	LPRIALR	.0011 x$(480/3800)^2$ = .00001755 H
Transformer secondary equivalent series leakage inductance per line	LSECAL	.000018 H
Transformer secondary equivalent series resistance per line	RSECA	.0078 Ω
Feeder line resistance	RL0A	.003 Ω
Feeder line inductance	LL0A	.00008 H
Branch line resistances	RL1A, RL2A, RL3A	.005 Ω
Branch line inductances	LL1A, LL2A, LL3A	.0013 H
Motor M1 line-to-neutral resistance	RM1A	3.2 Ω
Motor M1 line-to-neutral inductance	LM1A	.0041 H
Motor M2 line-to-neutral resistance	RM2A	4.3 Ω
Motor M2 line-to-neutral inductance	LM2A	.0055 H
Electronic load line-to-neutral resistance	REA	3.9 Ω
Electronic load line-to-neutral impedance	LEA	.0055 H

3) Write the netlist program.

```
Figure 5-2-2-1.CIR LOAD FLOW WITH THE REFLECTED VALUE METHOD
* Finding steady-state voltages and currents on a balanced three-phase circuit
*
VAR 1 0 AC 391.9 0
RSAR 1 2 .003191
LSAR 2 3 .00001276
*
LMAR 3 0 .3989
RMAR 3 0 351
RPRIAR 3 4 .007818
LPRIALR 4 5 .00001755
LSECAL 5 7 .000018
RSECA 7 8 .0078
*
RL0A 8 10 .003
LL0A 10 11 .00008
*
RL1A 11 12 .005
LL1A 12 13 .0013
RM1A 13 14 3.2
LM1A 14 0 .0041
*
RL2A 11 15 .005
LL2A 15 16 .0013
RM2A 16 17 4.3
LM2A 17 0 .0055
*
RL3A 11 18 .005
LL3A 18 19 .0013
REA 19 20 3.9
LEA 20 0 .0055
*
.PRINT AC V(1,0) I(RSAR) V(5,0)
*
.PRINT AC I(RSECA) V(11,0)
*
.PRINT AC V(13,0) I(RL1A)
```

```
*
.PRINT  AC  V(16,0)  I(RL2A)
*
.PRINT  AC  V(19,0)  I(RL3A)
*
.AC  LIN  1  60  60
*
.END
```

4) Run the PSpice simulation. The AC data requested in the netlist is in the output file.

```
**** 06/17/08 17:43:09 ******* PSpice Lite (August 2007) ****** ID# 10813 ****

Figure 5-2-2-1.CIR LOAD FLOW WITH THE REFLECTED VALUE METHOD

****     AC ANALYSIS                TEMPERATURE =   27.000 DEG C

******************************************************************************

 FREQ       V(1,0)     I(RSAR)    V(5,0)

 6.000E+01  3.919E+02  2.599E+02  3.879E+02

 FREQ       I(RSECA)   V(11,0)

 6.000E+01  2.575E+02  3.804E+02

 FREQ       V(13,0)    I(RL1A)

 6.000E+01  3.561E+02  1.002E+02

 FREQ       V(16,0)    I(RL2A)

 6.000E+01  3.625E+02  7.593E+01

 FREQ       V(19,0)    I(RL3A)

 6.000E+01  3.597E+02  8.144E+01
```

Figure 5-2-2-2 Part of the output file from running the netlist of Figure 5-2-2-1.

5.3 STEADY-STATE BALANCED THREE-PHASE SHORT-CIRCUIT

PROBLEM: The circuit of Section 5.1 has a balanced three-phase fault (all three phases connected together) at the input to motor M2. What are the steady-state line-to-line voltages and line currents at the transformer and loads? Use the PSpice transformer equivalent circuit method.

PROCEDURE:

1) Draw the basic one-line diagram.

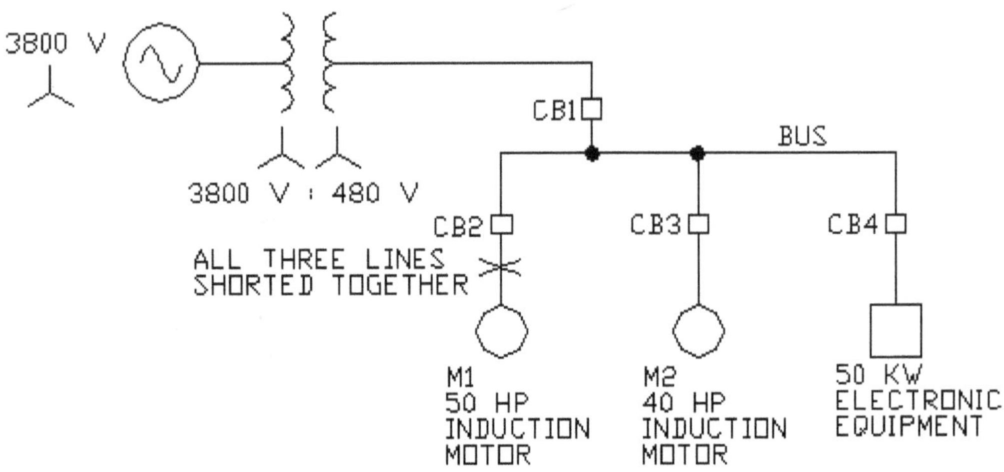

Figure 5-3-1 Balanced three-phase steady-state short-circuit one-line diagram.

2) As in Section 5.2, the circuit is balanced. So, it is only necessary to solve one line-to-neutral portion of the three-phase circuit. Only the A phase is drawn. It is drawn with equivalent line impedances, a transformer equivalent, motor and electronic equipment equivalent impedances, and node numbers.

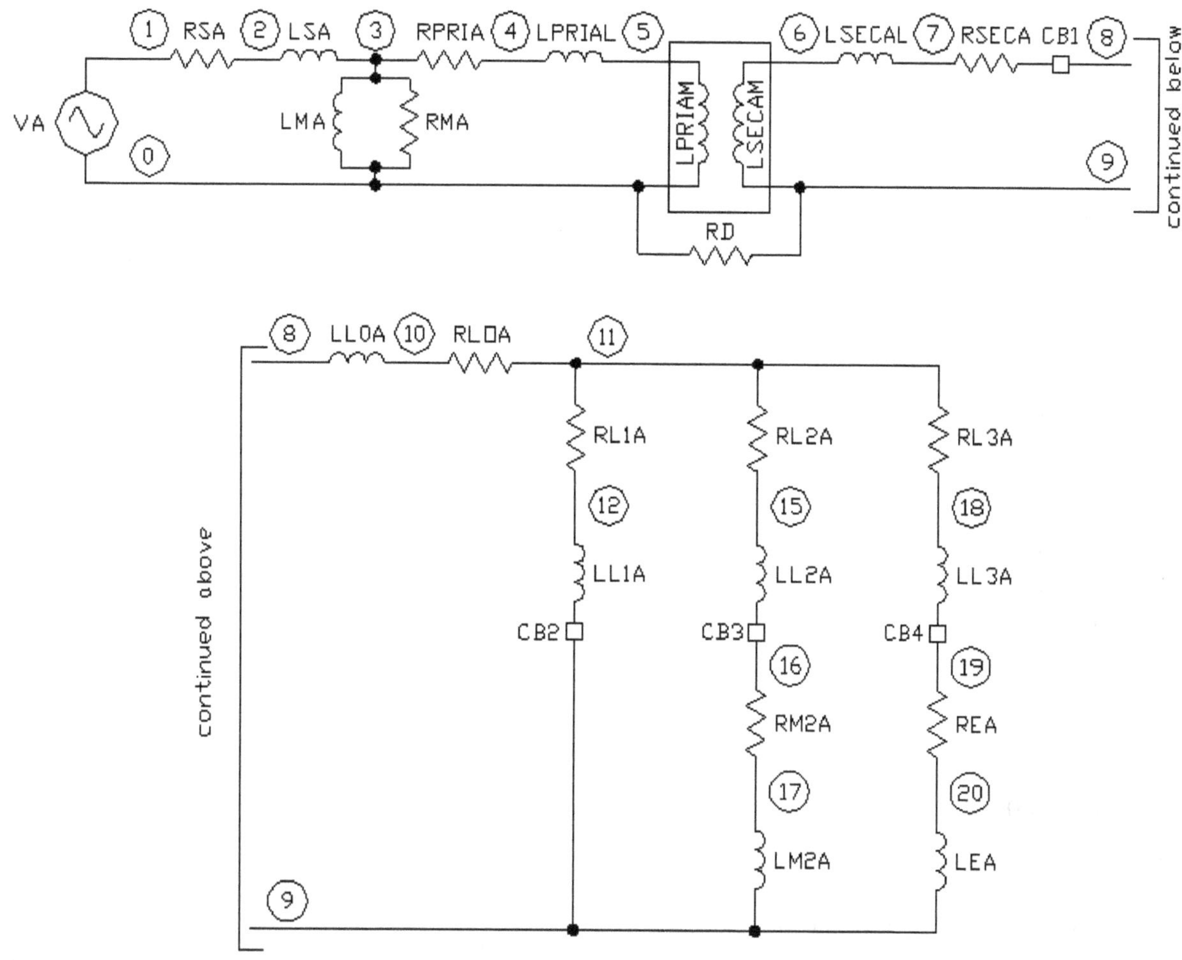

Figure 5-3-2 Line-to-neutral A phase equivalent circuit made ready for analysis by PSpice.

3) The netlist and variable values are the same as in Section 5.2.

4) Write the netlist program.

```
Figure 5-3-2.CIR SHORT-CIRCUIT WITH THE PSPICE TRANSFORMER EQUIVALENT
CIRCUIT
* Finding steady-state voltages and currents on a balanced three-phase short-circuit
*
VA 1 0 AC 3103 0
RSA 1 2 .2
LSA 2 3 .0008
*
LMA 3 0 25
RMA 3 0 22000
RPRIA 3 4 .49
```

```
LPRIAL 4 5 .0011
LPRIAM 5 0 7.14
LSECAM 6 9 .114
LSECAL 6 7 .000018
RSECA 7 8 .0078
KPRI-SECA LPRIAM LSECAM 1.00
RD 0 9 1E12
*
RL0A 8 10 .003
LL0A 10 11 .00008
*
RL1A 11 12 .005
LL1A 12 9 .0013
*
RL2A 11 15 .005
LL2A 15 16 .0013
RM2A 16 17 4.3
LM2A 17 9 .0055
*
RL3A 11 18 .005
LL3A 18 19 .0013
REA 19 20 3.9
LEA 20 9 .0055
*
.PRINT AC V(1,0) I(RSA) V(5,0)
*
.PRINT AC V(6,9) I(RSECA) V(11,9)
*
.PRINT AC I(RL1A)
*
.PRINT AC V(16,9) I(RL2A)
*
.PRINT AC V(19,9) I(RL3A)
*
.AC LIN 1 60 60
*
.END
```

5) Run the PSpice simulation. The AC data requested in the netlist is in the output file.

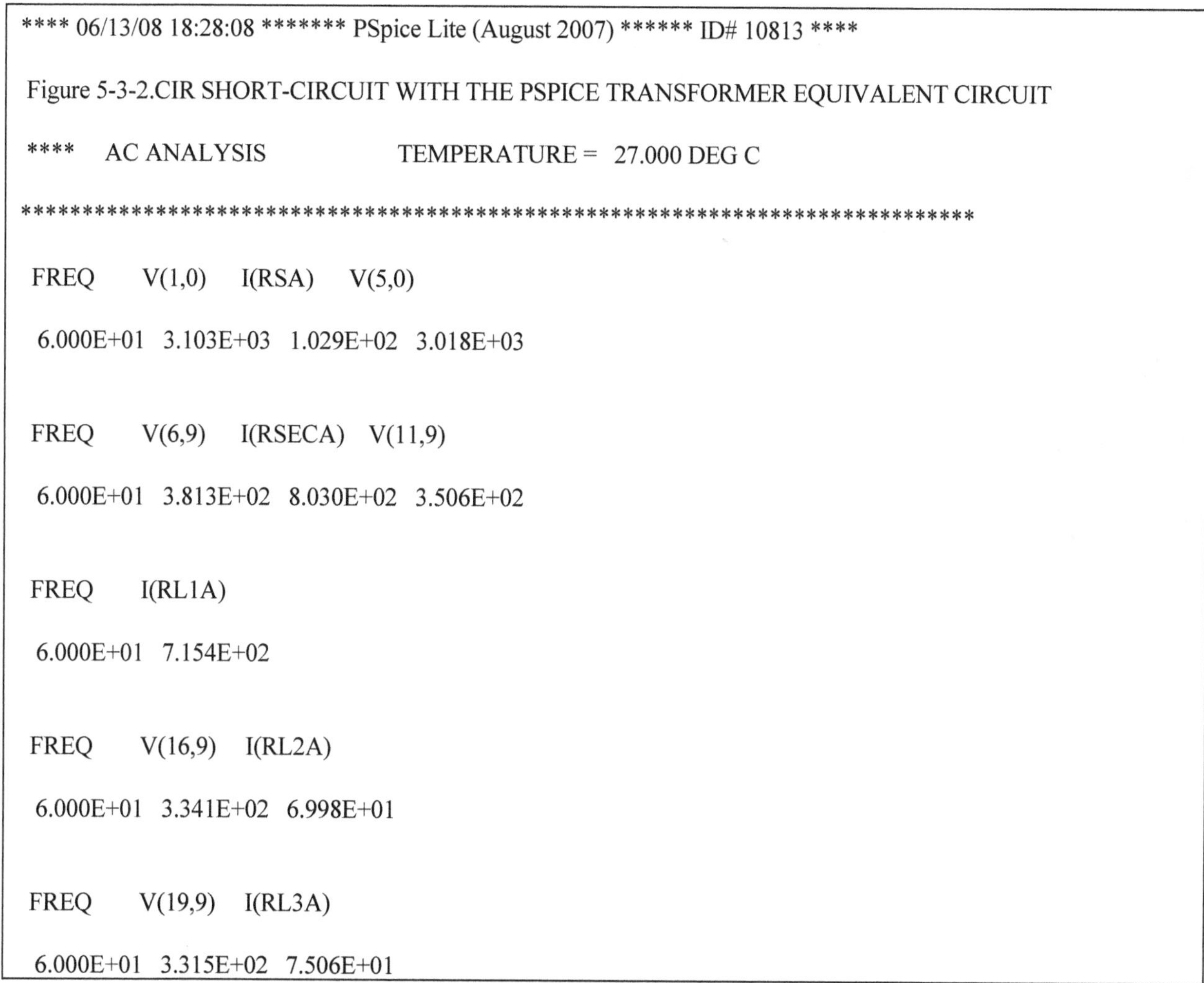

```
**** 06/13/08 18:28:08 ******* PSpice Lite (August 2007) ****** ID# 10813 ****

Figure 5-3-2.CIR SHORT-CIRCUIT WITH THE PSPICE TRANSFORMER EQUIVALENT CIRCUIT

****    AC ANALYSIS              TEMPERATURE =  27.000 DEG C

******************************************************************************

 FREQ       V(1,0)     I(RSA)     V(5,0)

 6.000E+01  3.103E+03  1.029E+02  3.018E+03


 FREQ       V(6,9)     I(RSECA)   V(11,9)

 6.000E+01  3.813E+02  8.030E+02  3.506E+02


 FREQ       I(RL1A)

 6.000E+01  7.154E+02


 FREQ       V(16,9)    I(RL2A)

 6.000E+01  3.341E+02  6.998E+01


 FREQ       V(19,9)    I(RL3A)

 6.000E+01  3.315E+02  7.506E+01
```

Figure 5-3-3 Part of the output file from running the Figure 5-3-2 netlist.

5.4 STEADY-STATE THREE-PHASE LINE-TO-LINE SHORT-CIRCUIT

PROBLEM: The circuit of Section 5.2 has an A-line to B-line fault at the input to motor M1. What are the steady-state line-to-line voltages and line currents at the transformers and loads?

DISCUSSION:

Unlike the problems of Sections 5.2 and 5.3, this will be solved with PSpice analyzing all the phases. This will produce current and voltage values throughout the circuit in one analysis. The reflected value method will be used here.

PROCEDURE:

1) Draw the basic one-line diagram.

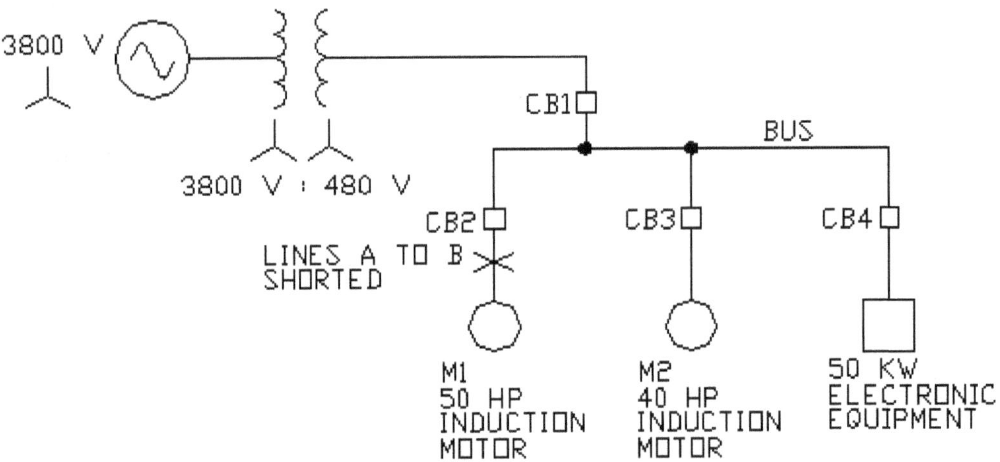

Figure 5-4-1 One-line diagram, steady-state line-to-line short-circuit.

2) Draw a PSpice equivalent circuit showing all three phases with the primary side voltage source, line impedances and transformer impedances reflected to the secondary side. Also include in the equivalent circuit the secondary side transformer impedances, line impedances, motor and electronic equipment equivalent impedances, and node numbers.

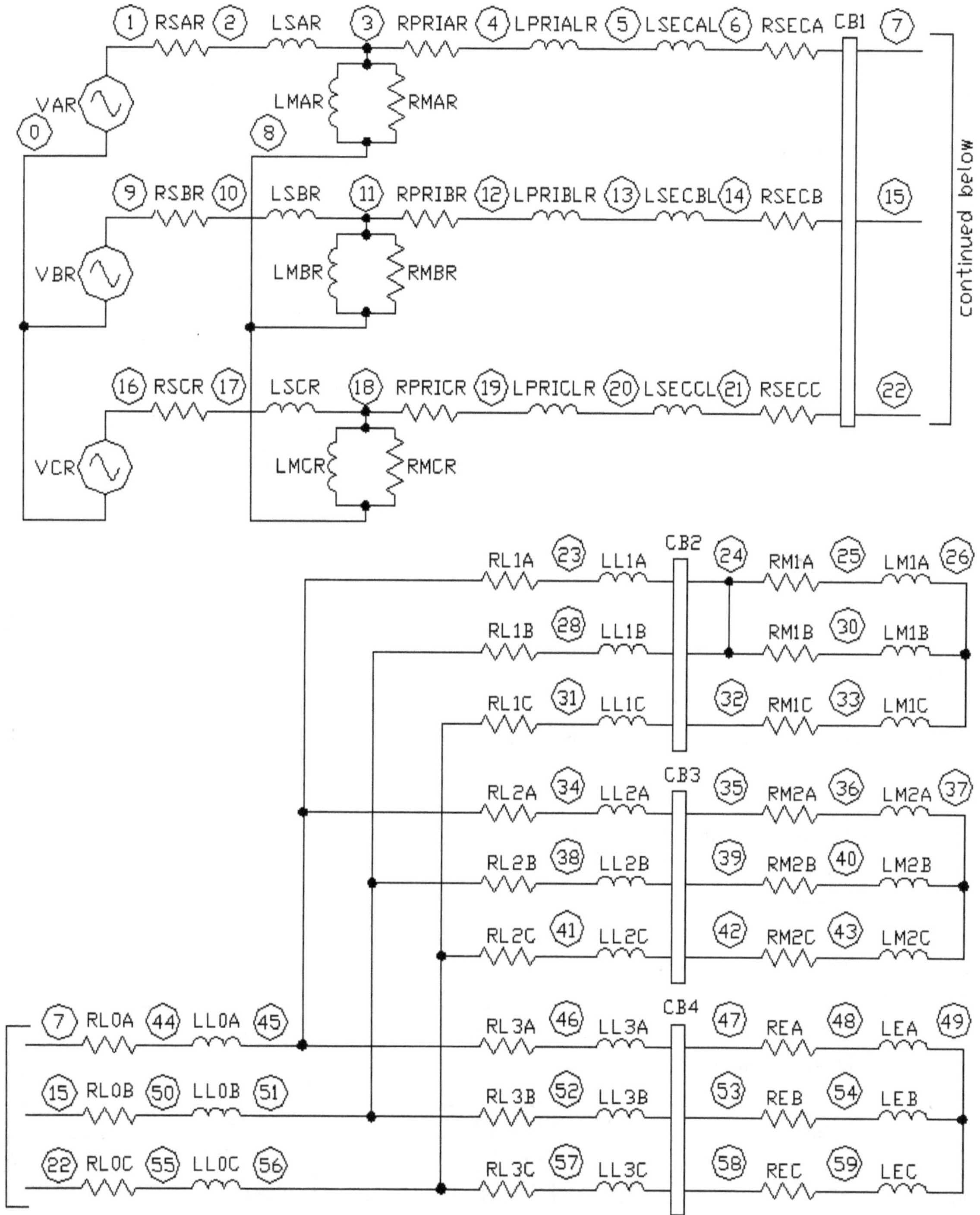

Figure 5-4-2 Steady-state A-line to B-line short-circuit ready for analysis by PSpice.

3) The netlist variable values are:

Description	Symbol	Value and units
Reflected A phase line-to-neutral AC input sinusoidal peak voltage	VAR	3103x(480/3800) = 391.9 V_{ACpeak}, 60 Hz frequency, 0° phase angle
Reflected B phase line-to-neutral AC input sinusoidal peak voltage	VBR	3103x(480/3800) = 391.9 V_{ACpeak}, 60 Hz frequency, 240° phase angle
Reflected C phase line-to-neutral AC input sinusoidal peak voltage	VCR	3103x(480/3800) = 391.9 V_{ACpeak}, 60 Hz frequency, 120° phase angle
Reflected system equivalent line resistances	RSAR, RSBR, RSCR	.2 $x(480/3800)^2$ = .003191 Ω
Reflected system equivalent line inductances	LSAR, LSBR, LSCR	.0008 $x(480/3800)^2$ = .00001276 H
Reflected transformer magnetizing inductances per phase	LMAR, LMBR, LMCR	25 $x(480/3800)^2$ = .3989 H
Reflected transformer magnetizing resistances per phase	RMAR, RMBR, RMCR	22000 $x(480/3800)^2$ = 351.0 Ω
Reflected transformer primary equivalent series resistances per line	RPRIAR, RPRIBR, RPRICR	.49 $x(480/3800)^2$ = .007818 Ω
Reflected transformer primary equivalent series leakage inductances per line	LPRIALR, LPRIBLR, LPRICLR	.0011 $x(480/3800)^2$ = .00001755 H
Transformer secondary equivalent series leakage inductances per line	LSECAL, LSECBL, LSECCL	.000018 H
Transformer secondary equivalent series resistances per line	RSECA, RSECB, RSECC	.0078 Ω
Feeder line resistances	RL0A, RL0B, RL0C	.003 Ω
Feeder line inductances	LL0A, LL0B, LL0C	.00008 H
Line resistances	RL1A, RL2A, RL3A, RL1B, RL2B, RL3B, RL1C, RL2C, RL3C	.005 Ω
Line inductances	LL1A, LL2A, LL3A, LL1B, LL2B, LL3B, LL1C, LL2C, LL3C	.0013 H

Continued

Motor M1 line-to-neutral resistances	RM1A, RM1B, RM1C	3.2 Ω
Motor M1 line-to-neutral inductances	LM1A, LM1B, LM1C	.0041 H
Motor M2 line-to-neutral resistances	RM2A, RM2B, RM2C	4.3 Ω
Motor M2 line-to-neutral inductances	LM2A, LM2B, LM2C	.0055 H
Electronic load line-to-neutral resistances	REA, REB, REC	3.9 Ω
Electronic load line-to-neutral impedances	LEA, LEB, LEC	.0055 H

4) Write the netlist program.

```
Figure 5-4-2.CIR LINE-TO-LINE FAULT WITH THE REFLECTED VALUE METHOD
* Finding steady-state voltages and currents on a three-phase line-to-line fault
*
VAR 1 0 AC 391.9 0
VBR 9 0 AC 391.9 240
VCR 16 0 AC 391.9 120
*
RSAR 1 2 .003191
RSBR 9 10 .003191
RSCR 16 17 .003191
LSAR 2 3 .00001276
LSBR 10 11 .00001276
LSCR 17 18 .00001276
*
LMAR 3 8 .3989
LMBR 11 8 .3989
LMCR 18 8 .3989
RMAR 3 8 351
RMBR 11 8 351
RMCR 18 8 351
```

```
*
RPRIAR 3 4 .007818
RPRIBR 11 12 .007818
RPRICR 18 19 .007818
*
LPRIALR 4 5 .00001755
LPRIBLR 12 13 .00001755
LPRICLR 19 20 .00001755
*
LSECAL 5 6 .000018
LSECBL 13 14 .000018
LSECCL 20 21 .000018
*
RSECA 6 7 .0078
RSECB 14 15 .0078
RSECC 21 22 .0078
*
RL0A 7 44 .003
RL0B 15 50 .003
RL0C 22 55 .003
LL0A 44 45 .00008
LL0B 50 51 .00008
LL0C 55 56 .00008
*
RL1A 45 23 .005
RL1B 51 28 .005
RL1C 56 31 .005
LL1A 23 24 .0013
LL1B 28 24 .0013
LL1C 31 32 .0013
*
RL2A 45 34 .005
RL2B 51 38 .005
RL2C 56 41 .005
LL2A 34 35 .0013
LL2B 38 39 .0013
LL2C 41 42 .0013
```

```
*
RL3A 45 46 .005
RL3B 51 52 .005
RL3C 56 57 .005
LL3A 46 47 .0013
LL3B 52 53 .0013
LL3C 57 58 .0013
*
RM1A 24 25 3.2
RM1B 24 30 3.2
RM1C 32 33 3.2
LM1A 25 26 .0041
LM1B 30 26 .0041
LM1C 33 26 .0041
*
RM2A 35 36 4.3
RM2B 39 40 4.3
RM2C 42 43 4.3
LM2A 36 37 .0055
LM2B 40 37 .0055
LM2C 43 37 .0055
*
REA 47 48 3.9
REB 53 54 3.9
REC 58 59 3.9
LEA 48 49 .0055
LEB 54 49 .0055
LEC 59 49 .0055
*
.PRINT AC V(1,0) V(9,0) V(16,0)
*
.PRINT AC V(5,13) V(13,20) V(20,5)
*
.PRINT AC I(RSAR) I(RSBR) I(RSCR)
*
.PRINT AC I(RL0A) I(RL0B) I(RL0C)
*
.PRINT AC V(45,51) V(51,56) V(56,45)
*
.PRINT AC I(RL1A) I(RL1B) I(RL1C)
*
.AC LIN 1 60 60
*
.END
```

5) The AC data portion of the output file:

```
**** 06/21/08 21:06:10 ******** PSpice Lite (August 2007) ******* ID# 10813 ****

Figure 5-4-2.CIR LINE-TO-LINE FAULT WITH THE REFLECTED VALUE METHOD

****     AC ANALYSIS                TEMPERATURE =  27.000 DEG C

******************************************************************************

 FREQ       V(1,0)     V(9,0)     V(16,0)

 6.000E+01  3.919E+02  3.919E+02  3.919E+02


 FREQ       V(5,13)    V(13,20)   V(20,5)

 6.000E+01  6.604E+02  6.749E+02  6.631E+02


 FREQ       I(RSAR)    I(RSBR)    I(RSCR)

 6.000E+01  7.954E+02  6.120E+02  2.599E+02


 FREQ       I(RL0A)    I(RL0B)    I(RL0C)

 6.000E+01  7.926E+02  6.097E+02  2.575E+02


 FREQ       V(45,51)   V(51,56)   V(56,45)

 6.000E+01  6.072E+02  6.589E+02  6.336E+02


 FREQ       I(RL1A)    I(RL1B)    I(RL1C)

 6.000E+01  6.608E+02  5.795E+02  1.002E+02
```

Figure 5-4-3 Part of the output file from running the netlist of Figure 5-4-2.

5.5 TRANSIENT THREE-PHASE LINE-TO-LINE SHORT-CIRCUIT

PROBLEM: The circuit of Section 5.2 has an A-line to B-line short-circuit at the input to motor M1. What are the transient currents through the M1 motor's circuit-breaker? What are the transient line-to-line bus voltages? What happens when the fault occurs at different times?

DISCUSSION:

The solution to this problem uses a steady-state induction motor equivalent circuit, a fixed resistance in series with a fixed inductance. This circuit was discussed at the beginning of Section 4.2. Since the motor's speed can be expected to change during short-circuit, the values of the equivalent resistance and inductance would actually change. There is also an effect of the motor feeding current into the fault that the steady-state equivalent circuit does not account for. The result is that the following PSpice solution cannot be relied on to be more than qualitatively correct.

PROCEDURE:

1) Draw the basic one-line diagram.

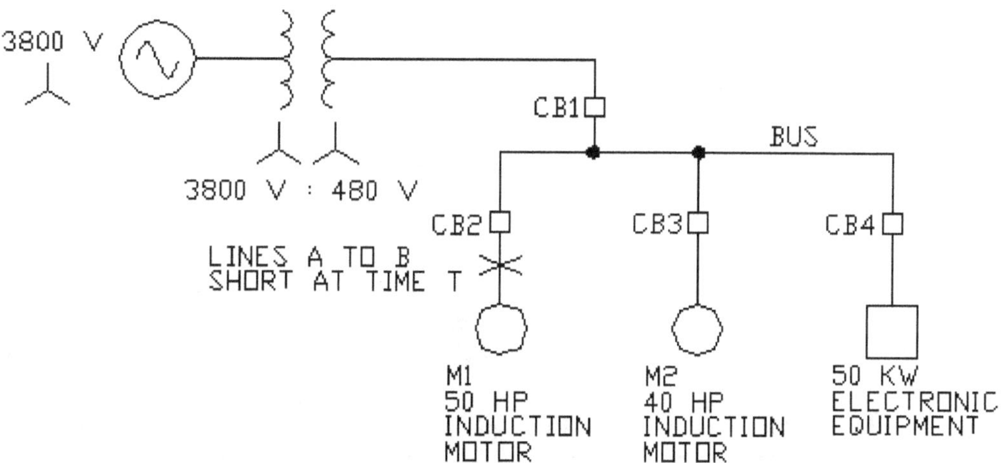

Figure 5-5-1 One-line diagram for a transient line-to-line short-circuit.

2) Draw a three-line diagram with impedances, a switch at the fault location, and node numbers. To make analysis easier all voltage and impedance values on the 3800 V side of the transformers will be reflected to the 480 V side.

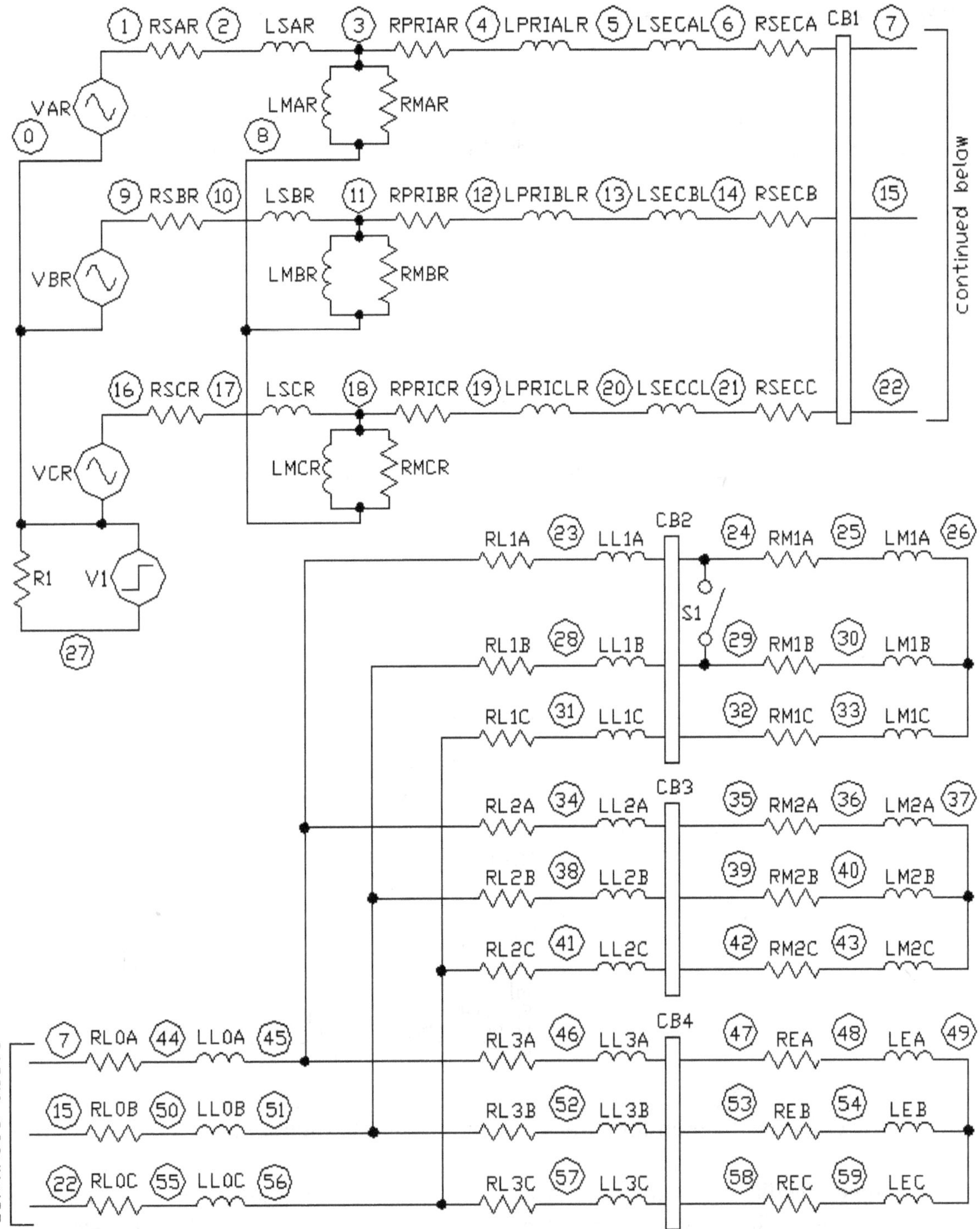

Figure 5-5-2 A-line to B-line short-circuit occurs at time T. The circuit is ready for analysis by PSpice.

3) Values for the supply voltage and components are:

Description	Symbol	Value and units
Reflected A phase line-to-neutral AC input sinusoidal voltage	VAR	0 V offset, 3103x(480/3800) = 391.9 V_{ACpeak}, 60 Hz frequency, 0 second delay time, 0 sec^{-1} damping factor, and 0° phase angle.
Reflected B phase line-to-neutral AC input sinusoidal voltage	VBR	0 V offset, 3103x(480/3800) = 391.9 V_{ACpeak}, 60 Hz frequency, 0 second delay time, 0 sec^{-1} damping factor, and 120° phase angle.
Reflected C phase line-to-neutral AC input sinusoidal voltage	VCR	0 V offset, 3103x(480/3800) = 391.9 V_{ACpeak}, 60 Hz frequency, 0 second delay time, 0 sec^{-1} damping factor, and 240° phase angle.
Reflected system equivalent line resistances	RSAR, RSBR, RSCR	$.2\ x(480/3800)^2 = .003191\ \Omega$
Reflected system equivalent line inductances	LSAR, LSBR, LSCR	$.0008\ x(480/3800)^2$ = .00001276 H, initial currents 0, 200, & -200 A
Reflected transformer magnetizing inductances per phase	LMAR, LMBR, LMCR	$25\ x(480/3800)^2 = .3989$ H
Reflected transformer magnetizing resistances per phase	RMAR, RMBR, RMCR	$22000\ x(480/3800)^2 = 351.0\ \Omega$
Reflected transformer primary equivalent series resistances per line	RPRIAR, RPRIBR, RPRICR	$.49\ x(480/3800)^2 = .007818\ \Omega$
Reflected transformer primary equivalent series leakage inductances per line	LPRIALR, LPRIBLR, LPRICLR	$.0011\ x(480/3800)^2$ = .00001755 H, initial currents 0, 200, & -200 A
Transformer secondary equivalent series leakage inductances per line	LSECAL, LSECBL, LSECCL	.000018 H, initial currents 0, 200, & -200 A

Continued

Transformer secondary equivalent series resistances per line	RSECA, RSECB, RSECC	.0078 Ω
Feeder line resistances	RL0A, RL0B, RL0C	.003 Ω
Feeder line inductances	LL0A, LL0B, LL0C	.00008 H, initial currents 0, 200, & -200 A
Line resistances	RL1A, RL2A, RL3A, RL1B, RL2B, RL3B, RL1C, RL2C, RL3C	.005 Ω
Line inductances	LL1A, LL2A, LL3A, LL1B, LL2B, LL3B, LL1C, LL2C, LL3C	.0013 H, initial currents 0, 70, & -70 A
Motor M1 line-to-neutral resistances	RM1A, RM1B, RM1C	3.2 Ω
Motor M1 line-to-neutral inductances	LM1A, LM1B, LM1C	.0041 H, initial currents 0, 70, & -70 A
Motor M2 line-to-neutral resistances	RM2A, RM2B, RM2C	4.3 Ω
Motor M2 line-to-neutral inductances	LM2A, LM2B, LM2C	.0055 H, initial currents 0, 70, & -70 A
Electronic load line-to-neutral resistances	REA, REB, REC	3.9 Ω
Electronic load line-to-neutral impedances	LEA, LEB, LEC	.0055 H, initial currents 0, 70, & -70 A
Switch circuit resistor	R1	1000 Ω
Switch	S1	On resistance = 1E-3 Ω, Off resistance = 1E6 Ω, Turn-on voltage = .6 V, Turn-off voltage = .4 V
Switch turn-on voltage pulse	V1	Voltage that starts to turn on at .0308 seconds and is totally on, at 1 volt, at .03081 seconds. In separate program runs .0353 and .03531 are used instead.

4) Write the netlist program. Note that approximate initial currents are written after the B and C lines' inductor values. In this circuit, if initial currents were not entered, PSpice would assume those initial currents were zero and then predict incorrect current waveforms.

```
Figure 5-5-2.CIR TRANSIENT LINE-TO-LINE FAULT WITH THE REFLECTED VALUE
METHOD
* Plotting circuit-breaker currents and line-to-line voltages versus time
* for different fault initiation times
*
VAR 1 0 SIN(0 391.9 60 0 0 0)
VBR 9 0 SIN(0 391.9 60 0 0 120)
VCR 16 0 SIN(0 391.9 60 0 0 240)
*
RSAR 1 2 .003191
RSBR 9 10 .003191
RSCR 16 17 .003191
LSAR 2 3 .00001276
LSBR 10 11 .00001276 IC=200
LSCR 17 18 .00001276 IC=-200
*
LMAR 3 8 .3989
LMBR 11 8 .3989
LMCR 18 8 .3989
RMAR 3 8 351
RMBR 11 8 351
RMCR 18 8 351
*
RPRIAR 3 4 .007818
RPRIBR 11 12 .007818
RPRICR 18 19 .007818
*
LPRIALR 4 5 .00001755
LPRIBLR 12 13 .00001755 IC=200
LPRICLR 19 20 .00001755 IC=-200
*
LSECAL 5 6 .000018
LSECBL 13 14 .000018 IC=200
LSECCL 20 21 .000018 IC=-200
```

```
*
RSECA 6 7 .0078
RSECB 14 15 .0078
RSECC 21 22 .0078
*
RL0A 7 44 .003
RL0B 15 50 .003
RL0C 22 55 .003
LL0A 44 45 .00008
LL0B 50 51 .00008 IC = 200
LL0C 55 56 .00008 IC = -200
*
RL1A 45 23 .005
RL1B 51 28 .005
RL1C 56 31 .005
LL1A 23 24 .0013
LL1B 28 29 .0013 IC = 70
LL1C 31 32 .0013 IC = -70
*
RL2A 45 34 .005
RL2B 51 38 .005
RL2C 56 41 .005
LL2A 34 35 .0013
LL2B 38 39 .0013 IC = 70
LL2C 41 42 .0013 IC = -70
*
RL3A 45 46 .005
RL3B 51 52 .005
RL3C 56 57 .005
LL3A 46 47 .0013
LL3B 52 53 .0013 IC = 70
LL3C 57 58 .0013 IC = -70
*
RM1A 24 25 3.2
RM1B 29 30 3.2
RM1C 32 33 3.2
LM1A 25 26 .0041
LM1B 30 26 .0041 IC = 70
LM1C 33 26 .0041 IC = -70
```

```
*
RM2A 35 36 4.3
RM2B 39 40 4.3
RM2C 42 43 4.3
LM2A 36 37 .0055
LM2B 40 37 .0055 IC = 70
LM2C 43 37 .0055 IC = -70
*
REA 47 48 3.9
REB 53 54 3.9
REC 58 59 3.9
LEA 48 49 .0055
LEB 54 49 .0055 IC = 70
LEC 59 49 .0055 IC = -70
*
* Time to short-circuit is chosen as .0308 and .0353 seconds in different runs
V1 27 0 PWL(0 0 .0308 0 .03081 1)
R1 0 27 1000
*
S1 24 29 27 0 RELAY
*
.MODEL RELAY VSWITCH(RON=1E-3 ROFF=1E6 VON=.6 VOFF=.4)
*
* Analysis times of .07 and .3 seconds are chosen in different runs
.TRAN .000005 .07 UIC
*
.PROBE
.END
```

5) Once the simulation is run, Probe can be used to display traces of voltages and currents throughout the circuit.

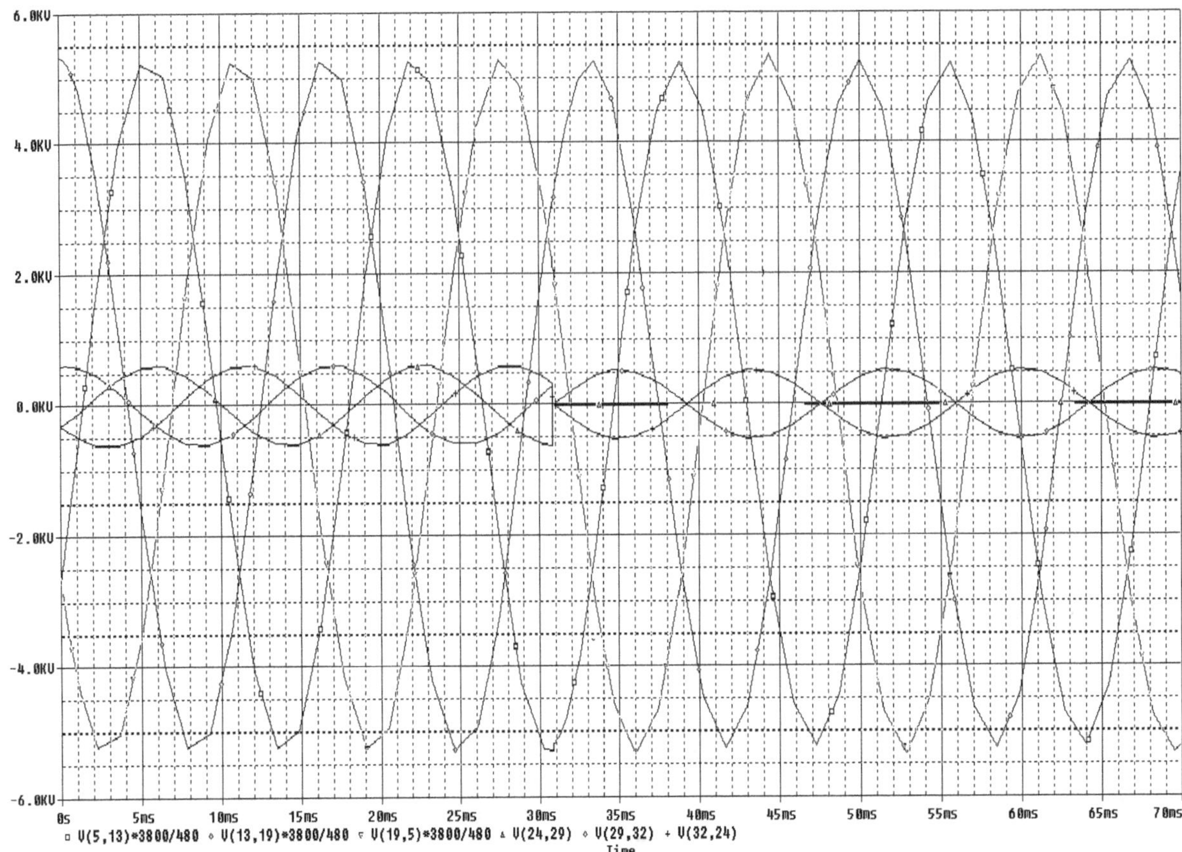

Figure 5-5-3 Line-to-line voltages input to the transformer and line-to-line voltages at the 480 V bus. Note that the transformer input voltages have been multiplied by (3800/480) in Probe to give them actual magnitudes. In this figure, the short circuit occurred at .0308 seconds.

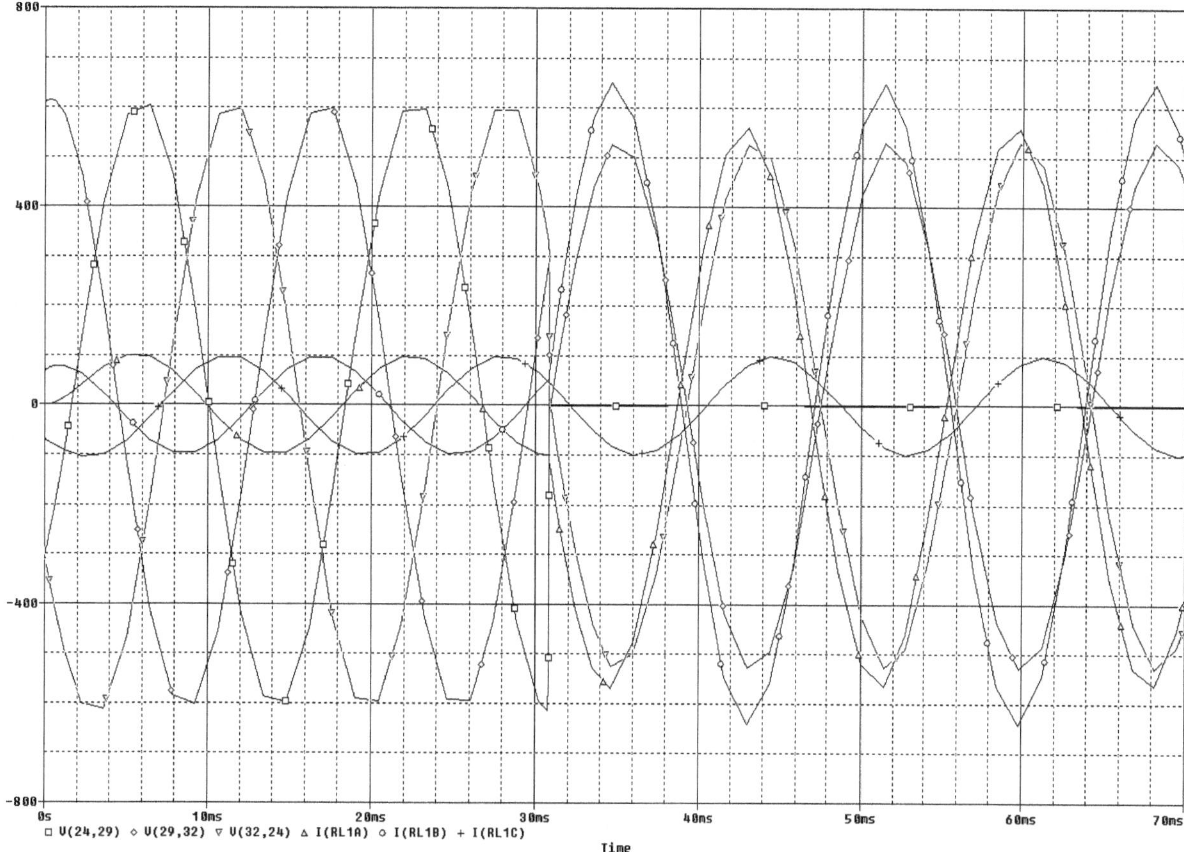

Figure 5-5-4 Line-to-line voltages at the 480 V bus and current into the M1 motor and then into the short-circuit. In this figure, the short-circuit occurred at .0308 seconds.

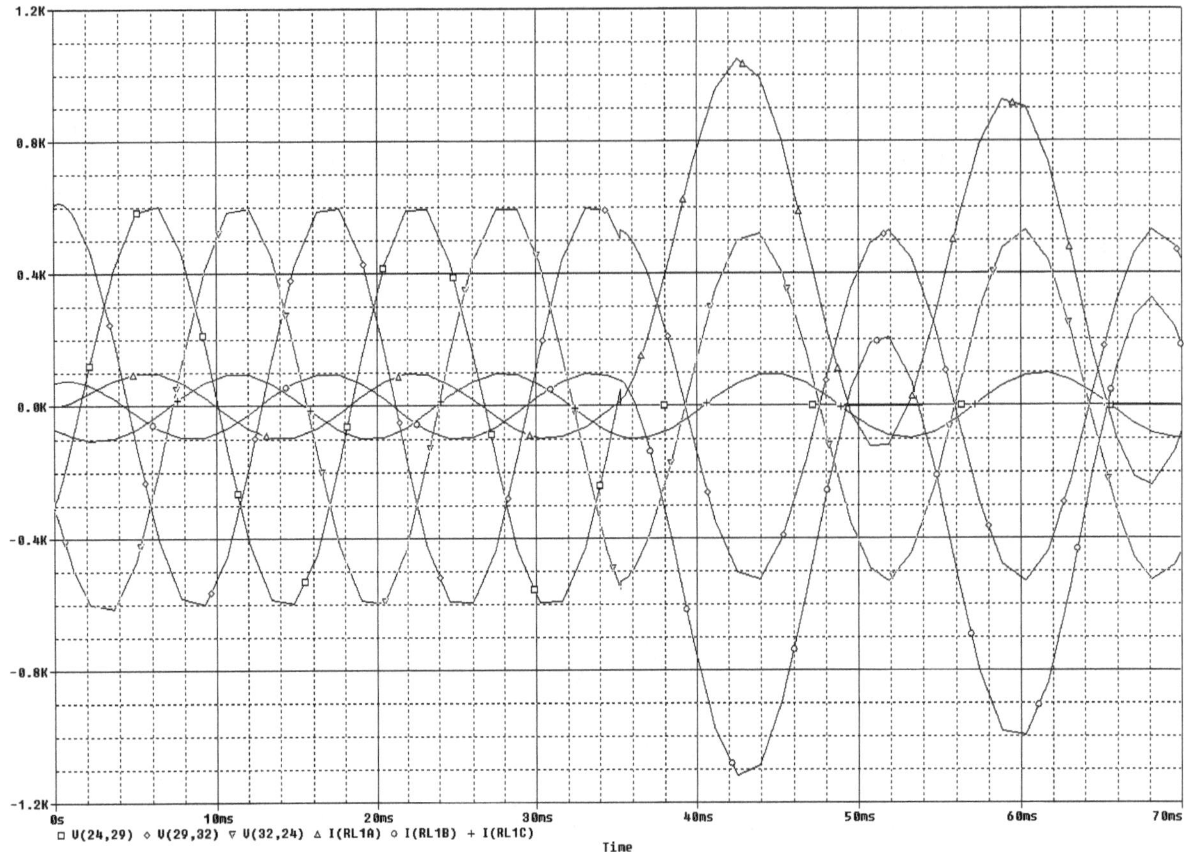

Figure 5-5-5 Line-to-line voltages at the 480 V bus and current into the M1 motor and then into the short-circuit. In this figure, the short-circuit occurred at .0353 seconds.

6) Notice how the initial short-circuit current is twice as high when the short-circuit occurs at .0353 seconds rather than .0308 seconds

7) Notice that the current going into the shorted phase has a decaying DC component and an AC component. The DC component might cause DC saturation problems in the transformers. These components can be seen in Figure 5-5-5, but can be more clearly seen in Figure 5-5-6 where Probe displays the rms and average features over a .3 second period.

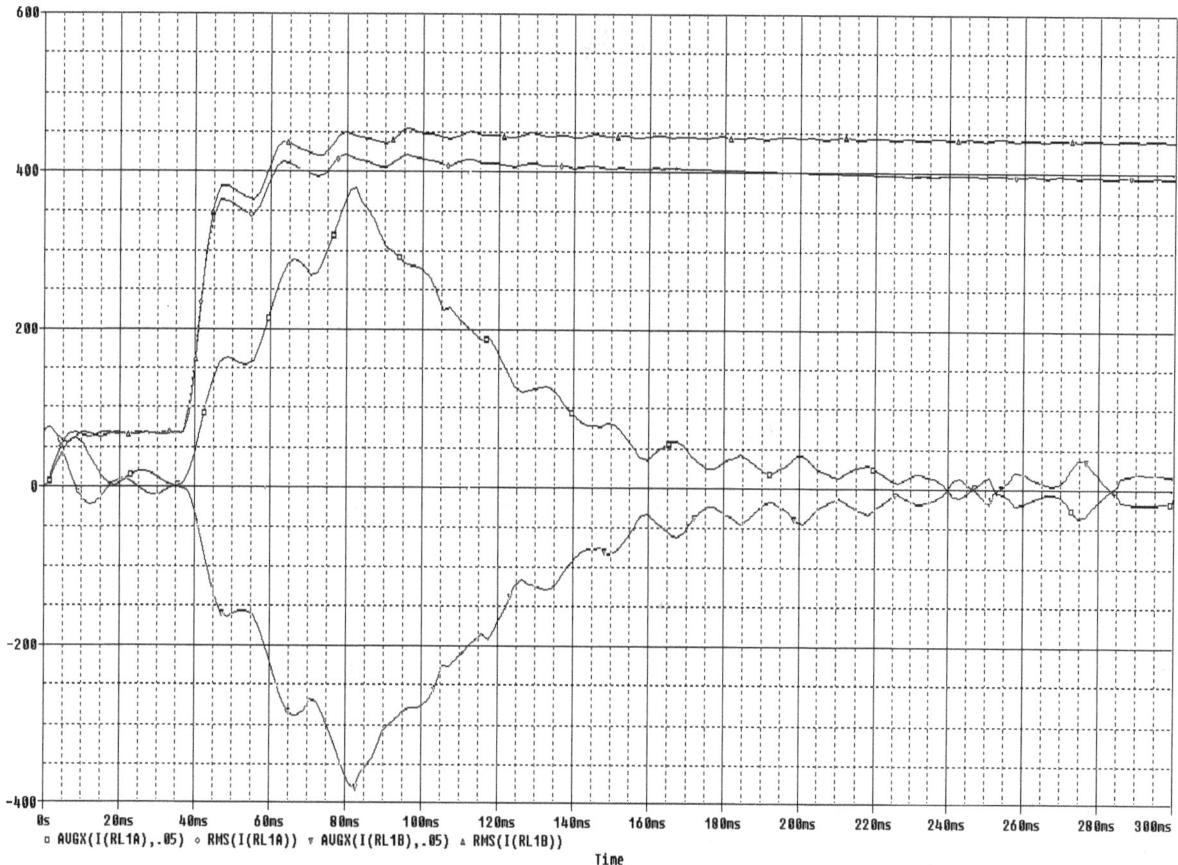

Figure 5-5-6 A and B phase rms and average currents going into the M1 motor and then into the short-circuit. In this figure, the short-circuit occurred at .0353 seconds.

8) If desired, the problem could be analyzed further.

a) The decaying DC component may drive the transformer into saturation in the time period immediately after short-circuit. More accurate transformer and motor models would produce more realistic results.

b) After short-circuit, the opening of the M1 motor circuit-breaker could be simulated. Switches and arc resistance simulating resistors could be put into the circuit to simulate the circuit-breaker opening. Since arc resistance is variable and difficult to predict, this analysis would be difficult.

6.0 OTHER SOURCES OF INFORMATION

6.1 USEFUL REFERENCES

Banzhaf, Walter. *Computer-Aided Circuit Analysis Using PSpice*, 2nd Ed. Englewood Cliffs, NJ: Regents/Prentice Hall, 1992, out of print.
This 7" x 9" book is 320 pages long. Banzhaf's book uses netlist programming exclusively. Basic netlist programming method has not changed, so the book is still relevant. Used copies are available from internet used book sellers.

Cadence. *PSpice User's Guide* Product Release 16.0, August 2007, included for free with the Demo CD.
This is the ultimate PSpice reference. It is not light reading. Unfortunately, Orcad Capture is used throughout, with no mention of netlist circuit entry.

eCircuit Center is a website at http://www.ecircuitcenter.com/.
It gives useful general information on various Spice programs including PSpice.

Mohan, N., Undeland, T., and Robbins, W. *Power Electronics: Converters, Application and Design*, Hoboken, NJ: John Wiley & Sons, 2003, price $118.64.
This 7" x 10" book is 824 pages long. It is a textbook with homework assignment problems. The book thoroughly covers power electronics and also provides PSpice analysis examples. It is a good book.

Ramshaw, R. and Schuurman, D. *PSpice Simulation of Power Electronics Circuits*, London, Chapman & Hall, 1996, price $185.00.
Ramshaw and Schuurman's book is similar to this one. The 6" x 9" book is 400 pages long. Their book contains many examples, all done with netlist rather than schematic capture programming. It is a good book, although it is pricey.

Rashid, Muhammad H. *Spice for Power Electronics and Electric Power*, 2nd Ed. Boca Raton, FL: CRC Press, 2005, price $119.95.
Rashid's book is similar to this one. The 6" x 9" book is 552 pages long. It would be a worthwhile purchase for a person who does many Spice simulations.

Tront, Joseph G. *PSpice for Basic Circuit Analysis*, 2nd Ed. New York, NY: McGraw Hill, 2005, price $34.06.
This 7" x 9" book is 141 pages long. Tront's book introduces students to the fundamentals of PSpice. Most of the book uses the Orcad Schematic Capture method to enter circuit data into PSpice rather than the netlist method. It would be a worthwhile purchase for the person who has little or no PSpice background and wants to learn the Schematic Capture method of circuit data entry.

6.2 CADENCE DESIGN SYSTEMS, INC.

EMA Design Automation
A Cadence Channel Partner
225 Tech Park Drive
Rochester, New York 14623
877-362-3321
www.ema-eda.com

EMA Design Automation has been contracted by Cadence to handle PSpice sales, service, and training.

7.0 APPENDIX

7.1 NETLIST STATEMENTS

This section only contains the statements used in this book. PSpice has many other statements.

INDEPENDENT VOLTAGE SOURCE STATEMENTS

- Vxxxxxx Node1 Node2 DC V

DC voltage for use without a DC voltage sweep.
Optionally can be used without "DC" as Vxxxxxx Node1 Node2 V
Node1 = Positive terminal
Node2 = Negative terminal
DC = DC
V = Voltage, volts
Used in Sections 4.2, 4.3.1, 4.3.2, & 4.4.

- Vxxxxxx Node1 Node2

DC voltage for use with a DC voltage sweep statement.
Node1 = Positive terminal
Node2 = Negative terminal
Vxxxxxx should also be specified in a .DC analysis statement. See ANALYSIS STATEMENTS.
Used in Section 3.1.

- Vxxxxxx Node1 Node2 AC V Phase

AC sinusoidal voltage for steady-state analysis used with an .AC analysis statement.
Node1 = Positive terminal at times when $(2\pi ft + \text{Phase angle [rad.]}) = \pi/2(\text{rad.})$
Node2 = Negative terminal at times when $(2\pi ft + \text{Phase angle [rad.]}) = \pi/2(\text{rad.})$
AC = AC
V = Peak voltage, volts
Phase = Phase angle, degrees
Used in Sections 3.2, 5.2.1, 5.2.2, 5.3, & 5.4.

• Vxxxxxx Node1 Node2 SIN(Vo Va Freq Td Df Phase)
AC sinusoidal voltage for transient analysis used with a .TRAN analysis statement.
Node1 = Positive terminal at times when ($2\pi ft$ + Phase angle [rad.]) = $\pi/2$(rad.)
Node2 = Negative terminal at times when ($2\pi ft$ + Phase angle [rad.]) = $\pi/2$(rad.)
SIN( ) = Sine wave
Vo = Offset voltage, volts
Va = Peak amplitude, volts
Freq = Frequency, Hz
Td = Time delay, seconds
Df = Damping factor, seconds^{-1}
Phase = Phase angle, degrees
Used in Sections 3.3.1, 4.1.1, 4.1.2, 4.3.1, & 5.5.

• Vxxxxxx Node1 Node2 PULSE(V1 V2 Td Tr Tf Pw Per)
Pulse voltage for transient analysis used with a .TRAN analysis statement. This repeats after completing each period, PER.
Node1 = Positive terminal
Node2 = Negative terminal
PULSE( ) = Pulse function, can produce square, triangle, and sawtooth waves.
V1 = Initial voltage, volts
V2 = Pulsed voltage, volts
Td = Time delay before pulse ramp starts, seconds
Tr = Rise time, seconds
Tf = Fall time, seconds
Pw = Time of pulse at pulsed voltage, seconds
Per = Period of pulse (time for waveform to repeat), seconds
Used in Section 3.3.2, 4.1.1, 4.1.2, 4.2, 4.3.1, 4.3.2, 4.4.

• Vxxxxxx Node1 Node2 PWL(T1 V1 T2 V2…up to…Tn Vn)
Piece-wise linear voltage for transient analysis with a .TRAN analysis statement.
The PWL function will produce any arbitrary voltage waveform. Unlike the PULSE function, the PWL function does not repeat after reaching its final time.
Node1 = Positive terminal
Node2 = Negative terminal
PWL( ) = Piece-wise linear function
V1 = Voltage at initial time, time 1, volts
T1 = Initial time of piece-wise linear voltage, seconds
V2 = Voltage at time 2, volts
T2 = Time 2 of piece-wise linear voltage, seconds
.
.
Vn = Voltage at last time, time N, volts
Tn = Last time, time N, of piece-wise linear voltage, seconds
Used in Section 5.5.

RLC ELEMENTS

• Rxxxxxxx Node1 Node2 R
Resistance
Node1 = Terminal 1
Node2 = Terminal 2
R = Resistance, Ω
Used in every circuit.

• Lxxxxxxx Node1 Node2 L IC
Inductance
Node1 = Terminal 1
Node2 = Terminal 2
L = Inductance, H
IC = Optional initial current at time zero, amps. If nothing is specified, PSpice will assume an initial condition of 0 A.
Used in most circuits.

• Cxxxxxxx Node1 Node2 C IC
Capacitance
Node1 = Terminal 1
Node2 = Terminal 2
C = Capacitance, F
IC = Optional initial voltage at time zero, volts. If nothing is specified, PSpice will assume an initial condition of 0 V.
Used in Sections 4.1.2, 4.2, 4.3.1, 4.3.2, & 4.4.

TRANSFORMER

• Kzzzzzzz Lxxxxxxx Lyyyyyyy Value
Kzzzzzzz = Designation of transformer
Lxxxxxxx = Designation of transformer primary mutual inductance, H
Lyyyyyyy = Designation of transformer secondary mutual inductance, H
Value = An ideal coefficient of coupling between the primary and secondary windings of 1.0 is usually used with power transformers.
Used in Sections 5.1.1, 5.2.1, & 5.3.

DIODE

• Dxxxxxx Node1 Node2 yyyyyyy
Dxxxxxx = Diode name
Node1 = Anode terminal
Node2 = Cathode terminal
yyyyyyy = Model type of the diode.
The model statement has the following format.
.MODEL yyyyyyy D(RS = rr BV = bv IBV =cb)
D( ) diode characteristics
rr = Diode forward resistance, Ω
bv = Reverse breakdown voltage, volts
cb = Current at breakdown voltage, amps
Used in Sections 4.1.1, 4.1.2, 4.2, 4.3.1, 4.3.2, 4.4.

ANALYSIS STATEMENTS

• .AC LIN No Fstart Fstop
AC analysis
LIN causes analysis to occur at evenly spaced frequencies from Fstart to Fstop.
No = Number of frequencies
Fstart = Start frequency, Hz
Fstop = Stop frequency, Hz
Used in Sections 3.2, 5.2.1, 5.2.2, 5.3, & 5.4.

• .DC Vxxxxxx Vstart Vstop Vstep
DC analysis
Vxxxxxx is the name of the DC voltage being swept
Vstart = Start DC voltage in the sweep, volts
Vstop = Stop DC voltage in the sweep, volts
Vstep = Step voltage in the sweep, volts
Used in Section 3.1.

• .TRAN Time step Time stop Time start Time computing step UIC
Transient analysis
Time step = Time between printed or plotted transient analyses when the .PLOT or .PRINT statements are used, seconds.
Time stop = Time at which analysis stops, seconds
Time start = Time at which analysis is first printed or plotted. The use of this is optional. If not stated, the start time is 0 seconds.
Time computing step = The largest computing step that PSpice will use, seconds. It is optional. If it is not specified, the largest computing step will be (Time stop – Time start)/50.
UIC = Use initial conditions statement. Initial conditions, IC, describe capacitor initial voltages and inductor initial currents. It is optional. If it is not specified, all initial conditions will be assumed to be 0.
Used in Sections 3.3.1, 3.3.2, 4.1.1, 4.1.2, 4.2, 4.3.1, 4.3.2, 4.4, 5.5.
Used with UIC in Section 5.5.

VOLTAGE CONTROLLED SWITCH

• Sxxxxxx Node1 Node2 Control node1 Control node2 Name
Node1 = Terminal 1
Node2 = Terminal 2
Control node1 = Positive terminal of the voltage that controls the switch
Control node2 = Negative terminal of the voltage that controls the switch
Name = Name of the switch. This refers to a .MODEL statement that describes the switch.
The Name statement has the following format:
.MODEL Name VSWITCH(RON=Ron ROFF=Roff VON=Von VOFF=Voff)
Name = Name of the switch
VSWITCH ( ) = Voltage controlled switch
Ron = Resistance of the switch when on, Ω
Roff = Resistance of the switch when off, Ω
Von = Voltage at which the switch has the Ron resistance, volts
Voff = Voltage at which the switch has the Roff resistance, volts
Used in Sections 4.1.2, 4.2, 4.3.1, 4.3.2, 4.4, & 5.5.

TABULATED OUTPUT

• .PRINT <analysis type> <variable1> <variable2>...
<analysis type> can be DC, AC, or TRAN
<variable_> = Variable to be printed
A statement of the analysis type needs to be in the netlist for .PRINT to work properly. See the ANALYSIS STATEMENTS above.
Used in Sections 3.1, 3.2, 5.2.1, 5.2.2, 5.3, & 5.4.

GRAPHED OUTPUT

• .PLOT <analysis type> <variable1> <variable2>...
<analysis type> can be DC, AC, or TRAN.
<variable_> = Variable to be printed and plotted over the range and step size specified by the analysis statement. See the ANALYSIS STATEMENTS above.
Used in Section 3.3.1.

• .PROBE <variable1> <variable2>...
<variable_> = Variable that Probe will graph versus time or frequency. Note, it is not necessary to put <variable_>'s after the .PROBE statement in the netlist. The variables can be entered after PSpice and Probe have been run by left clicking "Trace" and "Add Trace".
Also note that .PROBE has mathematical capabilities. See Section 7.2.
Used in Sections 3.3.1, 3.3.2, 4.1.1, 4.1.2, 4.2, 4.3.1, 4.3.2, 4.4, & 5.5.

DATA ENTRY

• .PARAM
Enters numerical data into a netlist. .PARAM is used with corresponding parameters and equations.
For example:
.PARAM F = 60
puts the value 60 in place of F in the netlist statement
V1 33 0 PULSE(0 100V {.04/F } 0 0 {.08/F} {1/F})
Used in Section 4.3.2 & 4.4.

• .END
End statement required at the end of every netlist.

7.2 PROBE MATHEMATICAL OPERATORS

Arithmetic operators: "+", "-", "*", "/", and "( )"
Trig Functions: "SIN(Y)", "COS(Y)", "TAN(Y)", and "ATAN(Y)"
Running Y average over the range of the x axis: "AVG(Y)"
Running Y average over the x axis from x-a to x: "AVGX(Y,a)"
Running Y RMS average over the range of the x axis: "RMS(Y)"
Maximum Y value over the range of the x axis: "MAX(Y)"
Minimum Y value over the range of the x axis: "MIN(Y)"
Square root Y: "SQRT(Y)"
Exponential power, Y^a: "PWR(Y,a)"
Log base e and base 10 of Y: "LOG(Y)" and "LOG10(Y)"
Derivative of Y with respect to the x axis: "d(Y)"
Integral of Y with respect to the x axis: "s(Y)"

7.3 LIMITATIONS OF PSPICE A/D DEMO

64 nodes
10 transistors
Two operational amplifiers
65 primitive devices
10 transmission lines (ideal or non-ideal)
4 pair-wise coupled lines
Device characterization using PSpice Model Editor of diodes only
Stimulus generation limited to sine waves (analog) and clocks (digital)
Sample library of 39 analog and 134 digital parts
Only displays simulation data made with the Demo version of the simulator
PSpice optimizer limited to one goal, one parameter, and one constant
Designs created in Capture can be saved if they have no more than 30 part instances.

7.4 MINIMUM HARDWARE REQUIREMENTS FOR PSPICE RELEASE 16.0

Intel Pentium 4 (32 bit) equivalent or faster
Windows Vista, Windows XP Professional (32 bit), or Window Server 2003
512 MB RAM (1G or more recommended for XP and Vista)
300 MB swap space
500 MB of free hard disk space if Capture is used
65,000 color Windows display with minimum 1024 x 768 (1280 x 1024) recommended
CD-ROM drive

www.ingramcontent.com/pod-product-compliance
Lightning Source LLC
Chambersburg PA
CBHW080423060326
40689CB00019B/4356

9780965944694